About the Author

Jerry Silver has developed components for terrestrial photovoltaic systems and designed solar arrays currently providing power for more than 20 commercial and NASA satellites. He participated in the production of high-performance semiconductor materials used for cell phone transistors, optical communication, and multijunction solar cells. Mr. Silver holds a B.S. in Engineering Physics from Cornell University and an M.S. in Physics from the University of Massachusetts. Mr. Silver currently teaches in the New Jersey area.

125 Physics Projects
for the Evil Genius

Evil Genius Series

Bike, Scooter, and Chopper Projects for the Evil Genius

Bionics for the Evil Genius: 25 Build-It-Yourself Projects

Electronic Circuits for the Evil Genius: 57 Lessons with Projects

Electronic Gadgets for the Evil Genius: 28 Build-It-Yourself Projects

Electronic Games for the Evil Genius

Electronic Sensors for the Evil Genius: 54 Electrifying Projects

50 Awesome Auto Projects for the Evil Genius

50 Green Projects for the Evil Genius

50 Model Rocket Projects for the Evil Genius

51 High-Tech Practical Jokes for the Evil Genius

Fuel Cell Projects for the Evil Genius

Mechatronics for the Evil Genius: 25 Build-It-Yourself Projects

MORE Electronic Gadgets for the Evil Genius: 40 NEW Build-It-Yourself Projects

101 Outer Space Projects for the Evil Genius

101 Spy Gadgets for the Evil Genius

123 PIC® Microcontroller Experiments for the Evil Genius

123 Robotics Experiments for the Evil Genius

PC Mods for the Evil Genius

Programming Video Games for the Evil Genius

Solar Energy Projects for the Evil Genius

Telephone Projects for the Evil Genius

22 Radio and Receiver Projects for the Evil Genius

25 Home Automation Projects for the Evil Genius

125 Physics Projects for the Evil Genius

JERRY SILVER

New York   Chicago   San Francisco   Lisbon
London   Madrid   Mexico City   Milan   New Delhi
San Juan   Seoul   Singapore   Sydney   Toronto

The McGraw·Hill Companies

Library of Congress Cataloging-in-Publication Data

Silver, Jerry.
  125 physics projects for the evil genius / Jerry Silver.
    p. cm.
  Includes index.
  ISBN 978-0-07-162131-1 (alk. paper)
  1. Physics—Experiments. I. Title. II. Title: One hundred twenty-five physics projects for the
evil genius.
  QC33.S5945 2009
  530.078—dc22

                                                                          2009002373

McGraw-Hill books are available at special quantity discounts to use as premiums and sales promotions, or for use in corporate training programs. To contact a representative please visit the Contact Us pages at www.mhprofessional.com.

125 Physics Projects for the Evil Genius

1 2 3 4 5 6 7 8 9 0 QPD/QPD 0 1 3 2 1 0 9

ISBN 978-0-07-162131-1
MHID 0-07-162131-8

Sponsoring Editor Judy Bass	**Proofreader** Paul Tyler
Acquisitions Coordinator Michael Mulcahy	**Indexer** Jerry Silver
Editing Supervisor David E. Fogarty	**Production Supervisor** Richard C. Ruzycka
Project Manager Patricia Wallenburg	**Composition** Patricia Wallenburg
Copy Editor Marcia Baker	**Art Director, Cover** Jeff Weeks

This book is for my wife Joanie and my kids Ally and Danny.

Acknowledgments

The author would like to gratefully acknowledge Steve Grabowski, Dan Silver, Danielle Buggé, Tracey Jameson, Michael Dershowitz, the Wallshes, Brookhaven Labs, John Kenney, and the folks at PASCO for assistance with the illustrations in this book. In addition, special thanks are offered to Steve Grabowski, Chis Aleo, Tiberiu Dragoiu, Robin Nolte, Tom Misniak, and Kim Feltre for enabling me to be part of a world where physics is appreciated, promoted, and shared on a daily basis.

Contents

Section 5   Energy/Momentum

Contents

Section 10   The Earth

Section 11   The Twentieth Century

Contents

Introduction

Who This Book Is Written For

This book has been written for anyone who is interested in, obsessed with, or simply mildly curious about exploring physics. The experiments in this book are intended to serve as a resource for teachers at all levels to use in planning laboratory activities for their classes and to get ideas for demonstrations. This book can also provide a way for anyone not necessarily directly involved with an academic physics class—including parents, scout leaders, and hobbyists—to pursue the world of physics as far as their interests take them. Young children—and those facilitating their education—will be able to appreciate many of these experiments on an intuitive level—perhaps one day to revisit them in greater depth.

If you are looking for science project ideas, you should be able to find something in these pages to work with. Students involved in a first-year high school or college physics class will find the overall sequence familiar and hopefully won't have too much trouble finding their way around.

I imagine that readers with a wide range of interests, backgrounds, and available resources will look through these pages for ideas about physics experiments. For this reason, I have written the projects/experiments to be accessible to readers in a number of different ways and on a variety of levels. Most of the experiments include a way to get started without requiring elaborate equipment.

How This Book Is Organized

Each section starts with a list of required items followed by step-by-step methods. Because there is often more than one way to do a project, various options are given to accommodate varying experience, available resources, and interest levels among readers.

The expected outcome for the experiments is given to help you interpret your experimental results. First you will find the most qualitative and intuitive insights, followed by increasingly detailed descriptions. For those who are interested (and *only* those who are interested) equations are provided to complete the explanation for why the experiments work. The reader is invited to pursue only as much or as little detail as they care to. This book is not intended to be a textbook on physics theory. I have tried, however, to help readers connect with the next step they might be ready to take. Just to be sure, each project has a conclusion that spells out the point of the experiment.

The World of Physics: Discovery and Rediscovery

On more than one occasion in the history of physics, the greatest advances have taken place at a time when the conventional wisdom of the day was that everything had already been discovered and all that was left to work out were the details. The hands-on approach presented here is to help the reader to (re-)discover physics directly. (All I ask, is in your acceptance speech for the Nobel Prize for physics, you reserve a few kind words for this book.)

There is typically more than one way to do many of these experiments. At least one procedure from start to finish is given for each experiment. But also a range of alternative approaches and extensions can be found for most of them. Hopefully this book will leave you with ideas, not only for how to do these experiments, but also for how to come up with experimental ideas of your own.

What You Will Need: Tool Bin/Parts List

The Basics

Each of the projects in this book has a specific parts list, called "What you need." Because different readers will have access to different types of equipment, alternative approaches are presented. A list of major suppliers is given in Appendix A at the end of this book.

The following are some of the items you may want to have handy. Most of these items are available in typical physics labs and many can be improvised.

- stopwatch—a one-tenth-second resolution is sufficient because it easily exceeds human reaction time.

- ring stands—many of the projects in this book involve supporting or holding other components of the apparatus. While this can be accomplished in other ways, having a basic set of ring stands with a few clamps gives you more time to focus on setting up the experiment.

- meterstick—most metersticks have millimeter markings. Metersticks often serve multiple functions in addition to measuring, such as holding lenses and mirrors. The thinner ones usually fit better with the supports used in optical experiments.

- tape measure—most of your work will be in meters. It is easy enough to convert feet to meters, but given the choice, a tape measure with metric divisions is preferable.

- ruler—with millimeter divisions.

- spring(s)—springs with varying degrees of stiffness to compare are useful. The best are ones that can be partially stretched by a reasonable weight.

- pulley—the less friction and the lowest mass, the better.

- string and rope—various kinds. You will want at least some thin strong string. Weaker string that can break plays a role in Project 24.

- mass set—a range of masses from 10 g to 1 kg (1000 g). They should have an attachment point from the top and, ideally, also from the bottom.

- spring scale—these come in various ranges, from a full scale reading of 2.5 newtons (255 g) to a full-scale reading of 50 newtons (5100 g). If you are doing demonstrations before a group, a large circular version of the scale with oversize lettering is the way to go. (Not to quibble, but weight is a *force* that is read in newtons and mass is a measure of an object's inertia, which is measured in grams. Physics purists definitely prefer weighing objects in newtons.) Most of our work will be in the System International (SI), which, to oversimplify, is a fancy new name for the metric system. Spring scales also come calibrated in pounds (and if you must, dynes, which almost no one uses today)—if you happen to have one, you can do the math. Conversion factors can be found in Appendix B.

- balance—low-end electronic balances have become much more affordable and can be purchased for less than $50. Other options include analog triple-beam scales or the more elaborate digital balances.

- wire—several meters of insulated wire, such as American wire gauge (AWG) 18, 20, or 22.

- jumper wires—jumper wires with various combinations of attachments make the electrical projects go a lot smoother. One termination is called a *banana plug*, which easily connects a circuit to a meter or a power supply. Another is spring loaded and grips onto an electrical connection called an *alligator clip*. (In the UK, some people refer to these "croc" clips. I am *not* making this up.)

- DC power supply (or batteries with a wire connection)—some projects require the capability to adjust the voltage. This requires an adjustable power supply, which can be purchased as a component. The power supply pictured in Figure I-1 (PASCO, part number SE-8828) costs less than $150, and enables you to do all the projects in this book that call for a DC power supply. Reasonably priced DC power supplies can also be purchased from Sargent-Welch, part number WLS-30972-81 or Flinn, part number AP5375.

  Some labs also have adjustable power supplies built into the lab benches. If your

DC power supply consists of a battery, the voltage can be made variable by using various combinations of resistors. Other options for a power supply include a hand-held generator that produces a DC voltage when a crank is manually turned (such as PASCO, part number EM-8090 or SE-8645, or Sargent-Welch, part number WL2420).

- batteries—C or D cell battery and holder. You will not need a battery if you have a DC power supply except for Project 113.

- electrical meters—the most useful and versatile meter is a multimeter. Multimeters perform the functions of ammeters, voltmeters, ohmmeters, and many can also be used as a digital thermometer. Multimeters can be acquired for less than $50, which in many cases is less expensive than stand-alone ammeters or voltmeters. Some projects require *both* an ammeter and a voltmeter, so for those projects, *two* multimeters are needed. If a multimeter is not available, a separate ammeter and voltmeter are needed. Figure I-2 shows a multimeter available from Sargent-Welch, part number WLS-30712-60 (also from PASCO, part number SE 9786A, or Frey, part number 15-531978-21) that works with all the projects in this book. One word of caution: the multimeter is much more

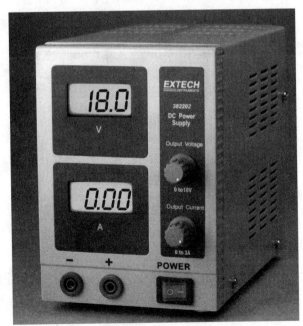

Figure I-1 *Power supply. Courtesy PASCO.*

Figure I-2 *Multimeter. Courtesy PASCO.*

versatile than a dedicated ammeter or voltmeter for most purposes. However, if used incorrectly, such as by being placed in series with too high a current, you can either blow a fuse, or worse, damage the meter.

- galvanometer—a galvanometer is a very sensitive ammeter that displays small electrical currents.

- electroscope—this is a simple device that measures the presence of static electric charges. They are inexpensive and available commercially. Homemade versions consist of a small ball attached by a wire. The wire is connected to two metal foil leaves, which are protected from discharge and air currents by a glass enclosure.

- magnets—bar and horseshoe magnets

- glassware—beakers, flasks

- rubber stoppers—two-hole and no-hole rubber stoppers to fit flasks

- hotplate—preferably adjustable with a ceramic top

- alcohol thermometer—mercury thermometers are no-no's in most labs today because of the environmental problems created if they break.

- hydrogen tube with high-voltage power supply

- calculator—very often the ideas and key insights in physics are revealed and discovered by doing a calculation. A simple scientific calculator, such as a TI-30 or equivalent, can be helpful in many of the projects in this book.

- Computers are used in many ways, including:
  - Collecting data from motion sensors and other measurement devices (such as light, sound, force, current, and voltage sensor).
  - Analyzing data in a spreadsheet, such as Excel, to identify mathematical models.
  - Sound card oscilloscope (see Project 64).

- laser pointer—a laser pointer with a replaceable (or, better yet, rechargeable) AA battery is the most versatile in the long run.

The simple laser pointer available in many dollar stores works well, but is less reliable.

- lenses—convex, concave, semicircular, rectangular, 45-degree, 60-degree, and right-angle prisms. Some lenses are "smoky," consisting of scattering particulates in the glass that make the light beams visible in the lens. Other lenses have a magnetic backing that makes them convenient to mount on a magnetic chalkboard or whiteboard. This makes it easy to enable ray tracing on a chalkboard. It is possible to glue a strong magnet to a lens, so they can be mounted on a chalkboard.

- mirrors—flat, concave, convex

- tape—duct tape, masking tape, electrical tape. One often-overlooked principle of physics is there is no such thing as having too much tape.

Things that are nice to have

- motion sensor—for less than the cost of a video game console, you can get a motion sensor that connects with your computor. Motion sensors (such as PASCO, part number PS-2103A) enable measurement of an object's position for various times. Figure I-3 shows the motion sensor. The Data Studio software that comes (free) with PASCO motion sensors lets you generate graphs of distance versus

Figure I-3 *Motion sensor. Courtesy PASCO.*

time, velocity versus time, and acceleration versus time with little prior experience with this equipment. The motion sensor requires a simple interface to the computer. The simplest of these approaches connects to the computer's USB port (PS-2100A) and requires no additional electrical power. Three sensors can be connected to a computer using the PS-2001.

- A hand-held data logger such as PASCO's Xplorer GLX (part number PS-2002) functions in a similar way with various sensors eliminating the need for a computer. This also enables measurements to be taken at more remote locations.

- oscilloscope—wave forms generated by sound picked up by a microphone or electrical signals can be displayed graphically on an oscilloscope. If you have one, you can do a number of interesting things with a scope. Because each oscilloscope is somewhat different, you also need a good manual or a patient friend to get you started. If you do not have a physical oscilloscope, you can inexpensively acquire software that can enable the sound card commonly available in computers to serve as a surprisingly functional oscilloscope. Details on how to do this can be found in Project 64.

- dry ice—the cloud chamber described in Project 125 uses dry ice, which can be obtained from welding supply companies, scientific labs, or chemical specialty companies. If you are able to get dry ice, you may want to get a little extra to explore other low-temperature physics experiments. Because dry ice, which is actually solidified carbon dioxide, is so cold, you should take precautions to avoid prolonged contact with the body. Use eye protection when working with dry ice, especially if you are breaking it into smaller pieces.

- Hover Puck—some physics labs use air tracks to eliminate friction. A lower-cost option is to use a Hover Puck that floats in a nearly frictionless manner across the floor. This is called out as an option in a few of the

experiments in this book. Hover Pucks are available from PASCO, part number SE-73358, and Kick It Stick It Disc from Sargent-Welch, part number WLS-1764-09.

- liquid nitrogen—liquid nitrogen is needed to make the ceramic material described in Project 106 cold enough to become superconducting. As with dry ice, liquid nitrogen is a material that is interesting in it own right and is explored in Project 92. It may make sense to plan both activities together. Just so you know: dry ice does *not* get cold enough to do Project 106 and liquid nitrogen is not recommended for Project 125 because it is too cold. Liquid nitrogen must be stored in a specially designed thermal container called a *liquid nitrogen dewar*, which safely handles the pressure that builds as the liquid nitrogen warms up. A regular thermos bottle with a sealed cap or any other type of sealed container should not be used. Liquid nitrogen is distributed in specially designed storage cylinders to organizations that do low-temperature studies, thermal cycling to test product reliability, and that use large volumes of gaseous nitrogen.

- vacuum pump—a vacuum pump is used in Projects 18, 41, and 94.

Wish-list

Not so many years ago, some of the greatest physics experiments remained the province of obscure physics labs in exclusive universities. Today, these experiments are within the reach of many physics departments with a moderate budget. Because the price tag for doing these experiments is thousands, rather than hundreds, of dollars, for most of us, they are considered here to be wish-list experiments. For each of these, a simpler, low-budget option is presented. The three wish-list experiments referred to in this book are:

- Millikan oil-drop experiment (PASCO, part number AP-8210, and Flinn Scientific, part number AP5671)

- photoelectric effect apparatus, such as the Daedelon EP-05 (available from www.daedelon.com or Flinn Scientific, part number AP5768).
- Cavendish gravitational constant (PASCO, part number AP-8215)

Yard sale physics

At the other end of the funding spectrum are items that can be adapted for use in physics experiments. As has been demonstrated by many of the great scientists of the past, much can be accomplished through resourcefulness and ingenuity. Besides the bargain hunters and antique dealers, physics enthusiasts can, on occasion, be observed looking for unnoticed treasure at yard sales where other people fail to see the true value. Here are some of the items you might want to add to your bag of tricks.

- bowling ball—a bowling ball makes a good pendulum mass that can also give one of the most accurate measurements of gravitational acceleration. A bowling ball can be used in Projects 19, 22, 26, or 66, if available. A heavy-duty screw eye can be anchored by taping into a pilot hole smaller than the diameter of the screw. Be careful and thoroughly test your mechanical connection before experimenting. Bowling balls can also come in handy in investigating collisions.
- swivel chair—you want one that rotates with as little friction as possible. This is useful in the conservation of angular momentum studies. Just the bottom part without the seat can be used for studying spinning objects.
- bathroom scale—this can be used to explore static equilibrium and torque.
- blow dryers—a blow dryer is a handy way to produce a reasonably steady air flow. This is used to explore Bernoulli's principle in Project 43.
- fish tanks—the ones with glass bottoms are especially useful for optical projects using

laser beams. A fish tank can be made into a cloud chamber in Project 125 or used as the container of a mousetrap-fission demonstration in Project 123.

- slide projectors—old slide projectors or moving projects can be good sources of light.
- laser levels—these can be used like a laser pointer. The beams are angled to produce a visible line along a wall, which can be advantageous for ray tracing. The output may not be the best-focused point source of light, so this is not the best choice, for instance, to use with a diffraction grating.
- turntables—turntables can be adapted for rotation experiments. (This can also be useful for the digital generation to see what a historic device like the phonograph looked like.)
- air hockey games—these work well with CDs as pucks and can be a good way to investigate elastic collisions.
- skateboards, rollerblades—to demonstrate Newton's third law.
- leaf blower—if you have a leaf blower, you have most of what you need to put together a one-person hovercraft. With just the right-shaped opening these can be used to levitate a beach ball in a demonstration of Bernoulli's principle (Project 43).
- bicycle tires—these make good gyroscopes and can be used for angular momentum experiments such as in Project 57.
- Christmas tree lights—these are an inexpensive and easy way to study electrical circuits such as in Project 100. They usually come in strands of series and parallel combinations.

You never know what else might come in handy, such as buckets, rope, wire, hotplates, clamps, lazy susans, golf balls, various tools, and motors. Keep your eyes open.

Motion

Project 1
Getting started. Constant velocity.
Running the gauntlet.

The Idea

With little or no friction to stop it, a moving object will keep moving at a constant velocity. This experiment explores a few simple ways you can take friction out of the picture.

What You Need

- Hover Puck
- tape measure
- 5 stopwatches
- masking tape
- several people to serve as timers

Method

1. Set up a course that is horizontal and free of obstructions. Do a trial run to make sure the Hover Puck does not move unless it's pushed and that it follows a reasonably straight line. (If you don't have a Hover Puck, a basketball or other similar object will do.)

2. Place distance markers, such as masking tape labels, at regular intervals. (Typically in physics, meters are used for distance. However, for this project any convenient unit can work as long as you're consistent throughout.)

3. Each of the timers should be assigned to measure the time at a specific distance along the path.

4. Timers should set their stopwatches to read zero and be prepared to start measuring the time as soon as the object starts moving.

5. Push the puck (or basketball) in the designated direction. Start with a *medium push*. See Figure 1-1.

6. As the puck (or basketball) passes each mark, each timer should stop the stopwatch and note the time.

7. Repeat with a slow push. A *slow push* is defined as slower than the medium push, but fast enough not to be pulled off course or stopped by friction.

8. Repeat with a *medium push*.

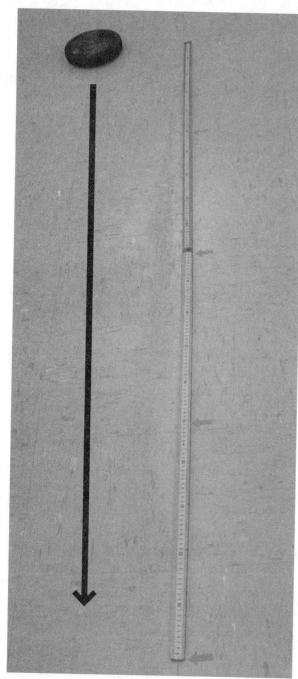

Figure 1-1 *Nearly frictionless motion can be achieved using a Hover Puck.*

9. Repeat with a *fast push*. This may be the most challenging one to time, especially for the first couple of timers.

10. The velocity for each of the runners will be the slope of the graph where distance is on the y-axis and time is on the x-axis.

Alternate Approach

Runners

1. Place the people with the timers on the 10, 20, 30, 40, and 50 yard lines of a football field.

2. Use a runner or several runners to run from the goal line to the 50 yard line.

3. As in number 2, get the time that each runner passes the designated distance marker, and then plot and interpret the results.

Expected Results

With constant velocity, each of the graphs should be *linear* (a straight line). The fastest runner has the highest slope, followed by the medium runner, with the slowest runner bringing up the rear.

If, for some reason, the motion was not perfectly constant, the points that differed will not be on the line. For instance, if the assumption that friction can be ignored is not completely valid, you may see some deceleration. In that case, the overall linear curve may be seen to taper off with a lower slope than the earlier points. If these data come from runners, it can be used to determine how steady the runners actually are. Also, if the runners start from zero, the first 10 yards will show an upward curve indicating acceleration.

Figure 1-2 shows expected results for three runs of 0.5, 1.0, and 1.5 meters per second (m/s).

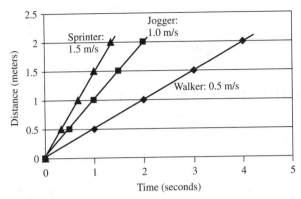

Figure 1-2 *Distance versus time for three different velocities.*

Figure 1-3 *Hovercraft. Courtesy PASCO.*

Why It Works

Average velocity can be thought of as the distance you go divided by the amount of time it took to get there. More specifically, we can say *average velocity* is the change in distance divided by the change in time. $v = \Delta d/\Delta t$ is the slope of the distance versus time graph. (Δ is the Greek letter delta, which means "change in.")

Other Things to Try

This experiment can be done using a person riding in style in a Hovercraft, as pictured in Figure 1-3. This can be done as an interesting way to do the previous experiment or just simply for the fun of doing it.

Because of the nearly frictionless motion, the person moves at constant velocity and makes a perfect object to measure at various speeds. You can purchase a Hovercraft (PASCO, part number ME 9838).

A Hovercraft can also be built by following these basic steps:

1. Drill a hole in the center of a 3-to-4 foot diameter piece of plywood.

2. Cut a hole halfway between the center and the edge just large enough to fit the end of a leaf blower.

3. Staple a plastic sheet to the bottom of the Hovercraft. Trim off the excess plastic.

4. Insert a bolt from the underside of the Hovercraft, through a plastic spacer (made from a plastic coffee-can lid). Attach the bolt through washers on the top and bottom, and then secure it with a nut.

5. Tape all the seals between the leaf blower and the plywood, and the plastic sheet and the plywood, to make them as airtight as possible.

6. Cut several approximately 2-inch diameter vent holes in the plastic sheet a few inches from the outer circumference of the plastic spacer.

7. With the leaf blower turned on, a cushion of air should enable a person to move with a minimum of friction.

Detailed plans can be found at http://amasci.com/amateur/hovercft.html.

The tendency of a moving object to keep moving is called *inertia*, which is addressed in Newton's first law. This is the subject of experiments that follow.

The Point

Constant velocity is represented by a straight line on a distance versus time graph. The slope of the line is equal to the average velocity.

Project 2
Picturing motion. Getting a move on.

The Idea

In the previous experiment, we worked with constant velocity in one direction and found that the motion was represented by simple graphs whose slopes were straight lines. Here, you study the motion of a person going forward and back, fast and slow. You also measure the effect of speeding up and slowing down. These graphs will take on a new dimension. In this experiment you use a motion sensor with display software to get a better feel for what different types of motion look like. Graphs are used to show where an object is at various times.

What You Need

- motion sensor
- appropriate computer interface for the motion sensor
- (roughly) 8 inch by 10 inch piece of cardboard

Method

Motion sensor (PASCO Easy Screen)

1. Attach a motion sensor to your computer. If you have a PASCO motion sensor, it is connected through the computer's USB port by way of a computer interface. Follow the specific details provided by the sensor's manufacturer.

2. If you are using the PASCO sensor, select the Easy Screen to get started. Four motion patterns will come up on the screen. Select one to start with. Press Run (when you are ready).

3. Hold the board facing the motion sensor. (See Figure 2-1.)

4. Position yourself so you start at a distance of 1 meter from the screen. On the computer screen, you see a visual indicator or your position as a function of time.

5. Adjust your position to match the pattern on the screen. (Note: you might be tempted to think that moving forward is positive, but this is not the case here. Moving backward results in *increasing* the distance between yourself and the motion sensor. As a result, for our purposes here, this is the *positive* direction.)

6. Repeat for each of the patterns available on the Easy Screen.

Figure 2-1 *Matching a pattern using a motion sensor. Courtesy PASCO.*

Expected Results

Figure 2-2 shows the result of someone moving backward and forward in such a way that they match the target motion pattern. This represents holding still for two seconds at 0.5 meters distance, then moving back at 2.2 m/s, and then holding still for another two seconds at a distance of 1.8 meters. The person doing the matching does not have to think about this, but only needs to look at the screen and move to fit the pattern.

Constant velocity in the positive direction (which in this case is defined as *away* from the motion sensor) is represented by a *straight line* on a distance versus time graph. *The faster the motion, the steeper the slope.*

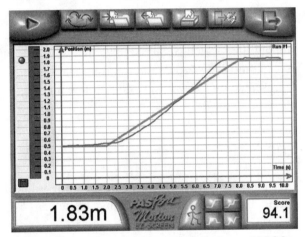

Figure 2-2 *Motion match results. Courtesy PASCO.*

Zero velocity means the distance stays the same over a given time interval. This is represented as a horizontal line on the distance versus time graph.

A curved line would be produced by accelerated motion (speeding up or slowing down).

Why It Works

The distance an object goes in a given time interval, t, is given by the equation:

$$d = d_o + vt$$

From this equation, the slope of the distance versus time graph is given by v, the velocity of the motion. The initial separation from the motion sensor, d_o, determines how far above the baseline the graph starts.

Each new phase of the motion contributes a separate segment to the graph. For instance, if the velocity stops, the distance remains constant for that period of time. If the motion is toward the motion sensor for another period of time, that motion contributes a segment of the graph with a negative slope that connects to the other segments.

Table 2-1 summarizes the various possibilities.

Table 2-1

		Overall slope	Shape of curve
Positive velocity (Away from the motion sensor)	Acceleration	Positive (Up to the right)	Upward
Positive velocity	Deceleration	Positive	Downward
Negative velocity (Toward the motion sensor)	Acceleration	Negative (Up to the left)	Upward
Negative velocity	Deceleration	Negative	Downward

Other Things to Try

A treasure map

1. On a piece of paper, draw the following moves (or make up your own):

   - Forward 1 meter in three seconds

   - In place four seconds

   - Back 0.5 meter in two seconds

   - Forward 2.5 meters in four seconds

   - In place four seconds

   - Back 1 meter in three seconds

2. How far did you get?

   - What was your *displacement?* (This is the *total distance you traveled from your starting point.*)

   - What was your average velocity? This is the *displacement divided by the total time* (including the time standing in place).

   - What was the total distance you traveled? Unlike displacement, every forward and backward move contributes to distance.

   - What is your *overall speed*? Your overall speed is the *distance divided by the total time*. The total time is the same for both of these.

The results of the treasure hunt is:

- Total time = 20 seconds

- Displacement from the starting point = +1 + 0 − 0.5 + 2.5 + 0 − 1 = 2.0 meters

- Average velocity = 2 meters / 20 seconds = 0.1 meters per second

- Total distance traveled = +1 + 0 + 0.5 + 1.5 + 0 + 1 = 4 meters

- Overall speed = 4 meters / 20 seconds = 0.2 meters per second

Make your own distance versus time challenges:

1. Select any of the Easy Screen Patterns.

2. Using a transparency marker (erasable or not is your choice) and trace the rectangular shape defining the Easy Screen Graph.

3. Draw your own motion pattern on the transparency.

4. Tape the transparency on the screen, so the rectangle aligns with the one you traced on the screen.

5. Match your pattern by adjusting your distance as before. This time, you will be ignoring the Easy Screen Pattern and following only your own.

Once you get the hang of it, you can throw in accelerated motion. Acceleration (away from the motion sensor) is represented by an upward sloping line, which is curved upward. Acceleration (toward the motion sensor) is represented by a downward sloping line that is curved downward.

The Point

Constant velocity is represented by a straight line on the distance versus time graph. The velocity is given by the slope of the line.

If the curve is not a straight line at any point this indicates that acceleration has occurred. Acceleration can be either positive (speeding up) or negative (called deceleration or slowing down).

An object moving in a particular direction (forward or backward) can experience either positive or negative acceleration.

Project 3
The tortoise and the hare. Playing catch-up.

The Idea

One car is going faster than the other, but the slower car has a head start. We can predict where and when the faster car will overtake the slower car. All we have to do is graph the movement of each car and see where the lines cross. This experiment gives you a method to make that prediction.

What You Need

- 2 toy cars with adjustable speeds
- stopwatch
- tape measure

Method

1. Set the speed of each of the two cars, so one is faster than the other. (If you don't know the speeds before starting, you can measure them in the following steps.)

2. Determine the average velocity of each of the cars by measuring the distance it goes in a given amount of time. The equation is average velocity = (distance traveled) divided by (time to get there). In physics, meters are typically used to measure distance (to be consistent with the *SI* or *System International* unit system). This will result in velocity measured in meters per second (m/s). However, you can work with other units for distance (such as feet per second) *as long as you are consistent.*

3. Line up the two cars in the same direction on a level floor heading in the same direction, as shown in Figure 3-1.

4. We are going to give the slower car a head start of a few seconds and try to predict where the faster car will overtake the slower car.

5. To do this:

   - Plot the speed of the faster car on a graph of distance versus time with the line starting at the origin and having a slope equal to the speed of the faster car.

   - Plot the speed of the slower car on the same graph, but starting at a point where the distance is zero and the time is equal to the chosen time delay.

   - See Figure 3-2, which shows a slower car going at 0.25 meter per second car given a 0.25 meter head start in front of a faster

Figure 3-1 *When will the faster car overtake the slower one?*

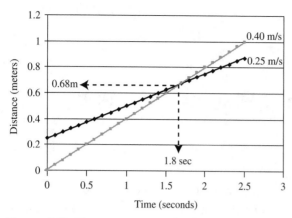

Figure 3-2 *Faster car passes the slower car where and when lines cross.*

car going 0.4 meter per second. (Notice the slower car is predicted to overtake the faster car at a point that is 0.68 meters from the starting point and 1.8 seconds after the race starts.)

6. Predict where the faster car you are working with will overtake the slower car.

7. Start the slower car and give it a head start.

8. Compare where and when the faster car will overtake the slower car with your predictions.

Expected Results

The faster car will overtake the slower car when the two lines in the graph cross. The distance the lines cross at is how far from the starting line the faster car catches the slower car.

The time where the lines cross is how many seconds from the start of the race when the slower car catches the faster car.

Why It Works

The distance that a object goes is given by the equation:

$$d = d_o + v(t - t_o)$$

where d_o is the initial distance between where the object starts and the starting line. (d_o can be understood as the head start in distance)

v is the velocity of the car

t is the time it has been going from the start of the race, and t_o is the delay or the head start in seconds given to the other car.

Other Things to Try

Here are some alternative ways of doing this:

1. If you have two motion sensors, focus one on the faster car and the other on the slower car. This generates a similar curve as shown in Figure 3-2. If the cars are moving *away* from you a similar curve will be produced, except the slope will be positive.

2. Another way to establish two different velocities is to use objects rolling off two different slopes starting from different heights. The object starting from the higher starting point will be rolling on the table or floor with a higher velocity, with the velocity proportionate to the height difference. If the bottoms of each of the ramps are the same distance from the starting line, the slower rolling object can be given a few seconds head start. A similar prediction and comparison of results can be made as in the previous section.

3. If you happen to be associated with a FIRST robotics team, you may want to consider using last year's robot(s) for this experiment.

4. Another variation is to predict where and when two cars moving toward each other will meet.

The Point

Two objects that move independently can be represented by separate equations that represent the relationship between distance and time. These are two simultaneous equations, which can be solved graphically to find the time and distance that the faster object overtakes the slower object.

Project 4
How does a sailboat sail against the wind?
Components of force.

The Idea

It is not hard to understand how a good stiff wind blowing from behind a sailboat can move it along at a brisk pace in the water. But what about getting back home? How can a sailboat move (or *tack*) against the wind?

In this project, you discover how a sailboat moving *against* the wind can result in a force that pushes the sailboat *forward*. This gets to the idea of how a force in one direction can be broken down into separate component forces. Two methods are shown here. The first method uses a sail attached to a pulley on a string. The second method uses an air track for those readers who have access to one. After looking at these methods, you are encouraged to try one or both of these, or to come up with your own idea.

What You Need

Pulley and string:

- stiff piece of foamboard or cardboard to use as a sail
- (low-friction) pulley
- small mass with an attachment hook, approximately 20 g
- 1–2 meters of thin string
- attachment points (such as ring stands clamped to a lab table) to hold the string horizontally
- blow dryer or other source of air flow
- duct tape

Air track:

- air track
- glider
- attachment for the glider that can hold a "sail." A bumper, for instance, can be attached to the top of a glider to serve as a "mast."
- 1 CD (or a stiff sheet of cardboard)

Method

Pulley and string

1. Attach the string horizontally to two anchor points. The string should be taut and able to support a small weight without sagging.

2. Hang the pulley on the string.

3. Hang the weight on the pulley so the pulley is free to slide on the string.

4. Tape the foamboard or cardboard at an angle of about 20–30 degrees with respect to the direction of the string.

5. With the sailboat supported on the string, direct the blow dryer at the sail. The blow dryer should be at a slightly greater angle (with respect to the string) than the angle of the sail. If the air from the blow dryer is too strong, the sail may vibrate. If the angle is too small, the sail will be forced backward with the wind. However, under the right conditions, the force in the forward direction will be strong enough to propel the pulley against the wind, in a similar manner to a real sailboat. See Figure 4-1.

Air track

1. Level the air track. You can determine that the air track is level by observing the glider when the air track is activated. If the glider does not move in either direction under the force of gravity, then the track can be considered to be level.

2. Attach a fixture on the glider that can hold a flat object, such as a CD.

3. Place the CD in the holder and secure it at an angle of about 20–30 degrees with respect to the air track.

4. Direct the blow dryer at a slightly greater angle than the angle of the sail, and then observe its response. See Figures 4-2 and 4-3.

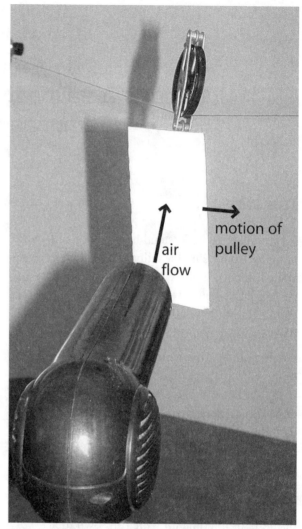

Figure 4-1 *At the correct angle, the blow dryer will draw the foamboard sail into the wind.*

Expected Results

For either method, the action of the blow dryer if positioned properly causes the "boat" to move *toward* the blow dryer. The boat is seen to move "against the wind." The parallel component of the force will cause the sailboat to move forward or tacking against the wind.

Using the pulley, if conditions are right, the perpendicular component of the force will also cause the sailboat to rotate around the string. This is comparable to a sailboat listing under the force of a strong wind. The keel of an actual sailboat serves to counteract the effect of this

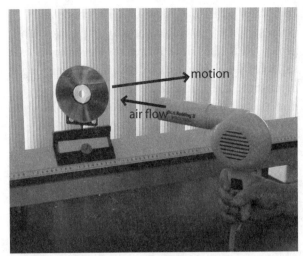

Figure 4-2 *Sailboat simulation using an air track viewed from the side. Photo by S. Grabowski.*

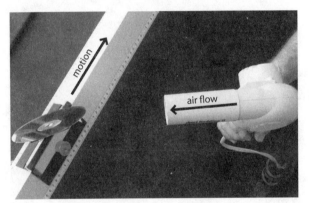

Figure 4-3 *Sailboat simulation using an air track viewed from the top. Photo by S. Grabowski.*

perpendicular force. In this experiment, this force is not constrained and causes the pulley to rotate.

Why It Works

The physical structure of a sailboat needs to do at least three things:

1. It picks up the force of the wind (roughly) perpendicular to the sail.

2. The keel of the sailboat makes the sailboat follow one-dimensional motion by preventing the sailboat from slipping perpendicular to its forward movement.

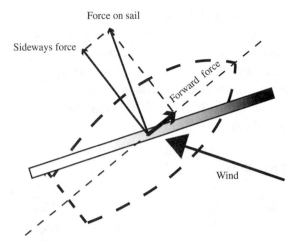

Figure 4-4 *Forces on a sailboat.*

3. It separates the force of the wind into two parts: one perpendicular to the movement of the boat, which is resisted by the keel, and one parallel to the motion of the boat, which propels it forward.

Figure 4-4 shows how the forces are separated into two components. The force produced on the sail by the wind blowing gets split up by the sailboat into two other forces. One tries to push the boat sideways and is resisted by the keel. The other force—if the angles are right—tries to push the boat forward. This happens even if the wind is coming more from in front than from behind. Quantities in physics that can be broken down into components as this force on a sailboat are called *vectors*.

Other Things to Try

Attaching a foamboard or cardboard sail to a toy car will work. The wheels of the car must turn freely and the tires must have enough friction to serve as a "keel" to restrict sideways motion.

Another way to do this is to use a (nearly) frictionless hockey puck with a low-friction tube to constrain motion in one dimension. A guide string (such as fishing line) is used to keep the motion in one dimension. You have to keep enough tension on the string to prevent the puck from rotating and binding. The puck must also be on a nearly perfectly horizontal surface. Tape a sail as in either of the two methods previously described. This approach also requires a reasonably horizontal surface to prevent the puck from drifting on its own before the blower is turned on.

The Point

A force in one direction can be thought of as being equivalent to two other forces pushing in completely different directions. This happens because force is a vector quantity in physics. This project illustrates how a force on the sail of a sailboat is the same as a sideways force pushing against the keel and a force in the forward direction of the sailboat. This is an example of the resolution of a force into two perpendicular components.

Project 5
Stepping on the gas.

The Idea

Pressing down with your foot on the accelerator of a car does not necessarily cause you to accelerate. You may be moving forward with constant velocity. How can you tell if you are accelerating? This experiment shows you a few ways to determine whether you are accelerating or just moving along at constant velocity.

In this project, you can also find ways to detect centripetal acceleration, which keeps things moving in a circle.

What You Need

Any or all of the following "accelerometers" can be used to detect acceleration:

- pendulum: any weight on a string
- float tied to a string held underwater
- candle
- partially filled tank of liquid
- accelerometer, such as shown in Figure 5-1

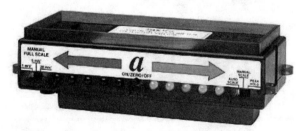

Figure 5-1 *Accelerometer. Courtesy PASCO.*

Method

Pendulum

1. Holding the string of the pendulum, move at as steady a pace as you can. Observe the pendulum during constant velocity.

2. Now do the same thing, but observe what happens when you speed up (accelerate).

Skateboard accelerometer

1. Attach a pendulum to a skateboard, as shown in Figure 5-2.

2. Roll it down a ramp that has a large enough slope for the skateboard to increase its speed. Observe the angle the pendulum makes with the vertical position.

3. Adjust the slope of the ramp, so the skateboard is just held on the ramp by friction without sliding down. This is called the *angle of repose*.

4. Give the skateboard a slight nudge. It should move at a fairly constant velocity. Note the angle of the pendulum.

5. What happens if the skateboard slows or goes up a ramp?

Centripetal acceleration

Spin an apparatus, such as shown in Figure 5-3 or 5-4. A pair of candles at either end of a spinning board is another way to do this. The floating bob apparatus is commercially available or can be assembled from fishing bobs (or Styrofoam

Figure 5-2 *"Skateboard" accelerometer.*

balls), baby food jars, a piece of wood with a hole in the center, and a metal post.

Expected Results

A pendulum hangs vertically when moving at constant velocity. But it moves in the *opposite* direction as the acceleration it is experiencing. If an object slows down or decelerates, it shows up as a backward movement in the pendulum.

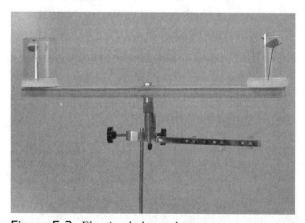

Figure 5-3 *Floating bob accelerometer.*

When the apparatus with the floating bob is spinning, the bob moves inward. This may be the opposite of what you might expect and is the opposite of what would happen with a freely hanging pendulum. The reason for this is the centripetal acceleration increases the buoyant force on the bob, forcing it inward. Candles move in the opposite direction. The flame moves outward, as does liquid in a container.

Why It Works

Newton's second law requires that force and acceleration are related to each other through F = ma. If there is acceleration (a), there is a force (F) on the moving object (or mass, m). The force is in the same direction as the acceleration.

Other Things to Try

An accelerometer, such as shown in Figure 5-4, directly indicates acceleration with a set of LEDs that light in proportion to the amount of acceleration. The greater the acceleration, the more LEDs will light. It can, for instance, indicate the acceleration of a cart pulled by a string. It can also be used to monitor centripetal acceleration.

The Point

A hanging (or other unconstrained) object is affected by acceleration, but is *not* affected by uniform steady velocity.

Figure 5-4 *An applied force causes an object to accelerate. Courtesy PASCO.*

Project 6
Rolling downhill. Measuring acceleration.

The Idea

When exposed to the force of gravity, objects fall faster and faster. This is called gravitational acceleration. When objects fall straight down, you have to be very quick if you want to measure how long an object falls a given distance. When Galileo Galilei tried to do this during the fifteenth century, he used primitive timing devices, such as dripping water and his own pulse to keep track of objects dropped from the Leaning Tower of Pisa. To overcome the difficulty of timing these measurements, Galileo had the brilliant insight of slowing down gravitational acceleration using a ramp. In this experiment, you follow in Galileo's footsteps. However, you have the advantage of being able to use a stopwatch or even a motion sensor to more accurately measure the object's movement.

What You Need

- inclined track (such as a section of wooden corner molding, semi-round vinyl bullnose molding, or a flat board with two "gutters" created by attaching metersticks as guides)
- golf balls or marbles
- stopwatch or other timer
- meterstick
- optional: motion sensor, inclined air track

Method

1. Set the inclined track at a moderate angle with respect to the surface on which it is supported.
2. Mark distance intervals from the bottom of the track in 10 cm increments.
3. Release the golf ball (or marble) from each of the distances marked and record the time in seconds that it takes to reach the end of the ramp. See Figure 6-1.

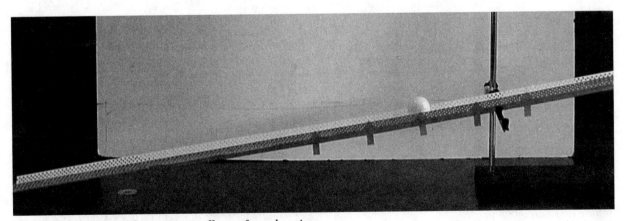

Figure 6-1 *Ramp used to measure effects of acceleration.*

4. If you measure the distance the ball rolls and the time it takes to roll, you can easily find the acceleration, a, at any point using $a = 2d/t^2$, where d is the distance it rolls and t is the time it takes to roll that distance.

5. What is the effect of changing the slope of the incline on the rate of acceleration?

Expected Results

A graph of distance versus time, such as pictured in Figure 6-2, shows that the distance the object moves *in a given amount of time* is increasing. The distance in the graph is shown to increase as the *square* of the time which is a characteristic of constant acceleration.

Why It Works

When an object accelerates, its velocity changes with time. For the case of constant acceleration, the velocity increases by a constant amount every second. This results in the distance increasing as the square of the time.

Other Things to Try

A rolling golf ball or marble can be considered a falling object whose acceleration is slowed by the incline. This is approximately, but not completely, true. Any rolling object develops angular

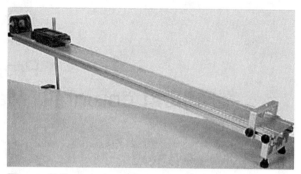

Figure 6-3 *An air track with a motion sensor attached to the end. Courtesy PASCO.*

momentum that ties up some of its energy in the process of rolling.

Better precision can be achieved by using an air track. This reduces the impact of friction and rotational kinetic energy. Incorporating a motion sensor to measure velocity and acceleration adds another dimension, as Figure 6-3 shows.

DataStudio software displays the distance measured by the motion sensor, as shown in Figure 6-4.

The Point

When an object accelerates, its velocity changes with time. If that acceleration is constant, distance increases as the square of time.

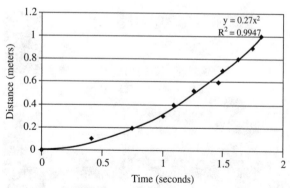

Figure 6-2 *Distance versus time for a golf ball rolling down an incline.*

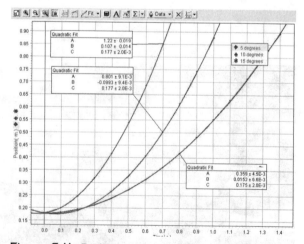

Figure 6-4 *Position (in meters) versus time (in seconds) for three different inclines. Courtesy PASCO.*

Project 7
Independence of horizontal and vertical motion. Basketball tossed from a rolling chair.

The Idea

Which will hit the ground first: a bullet dropped straight down from a height of 5 feet or a bullet fired horizontally over flat ground at 300 m/s from the same height? Many people guess that the greater momentum of the moving bullet would keep it in the air longer. This experiment addresses this question.

A *projectile* is an object that has both horizontal and vertical motion. Although motion in two dimensions may seem very complicated, it can be enormously simplified based on the results of this section. You discover that the horizontal motion of a projectile is completely independent of its vertical motion. It does not matter how fast an object is falling. In this experiment, you prove this in several ways.

What You Need

- chair
- basketball
- someone willing to sit in a chair
- independence of horizontal and vertical motion apparatus
- ballistic car

Method

Rolling chair

This is simple to do, but it has a significant result.

1. Have the person sit in the chair holding the basketball.
2. Roll the chair (with the person sitting). The person can alternatively be on a skateboard or roller blades.
3. Have the person toss the basketball up and observe its trajectory.

Coins

1. Place a coin at the edge of a table.
2. Flick a second coin toward the first so that the first is just pushed over the edge and the second coin flies off the table.
3. Both coins should start falling at the same time. One with a horizontal velocity and one without.
4. Listen to see which, if any, of the coins, strikes the floor first. Repeat this enough times until you get consistent results.

Apparatus

Use a commercially available apparatus, such as pictured in Figures 7-1 and 7-2. The apparatus shown in Figure 7-1 is much easier to use. The ballistics car shown in Figure 7-2 may require a level surface and some practice. A more reliable version of this is available as an accessory for an air track.

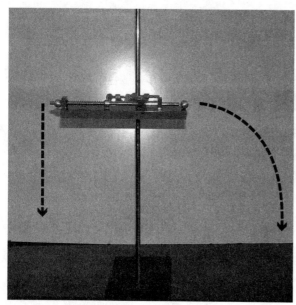

Figure 7-1 *Both balls hit the floor at the same time.*

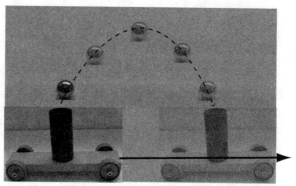

Figure 7-2 *Ballistics car showing the steel ball has the same horizontal velocity as the car.*

Expected Results

The basketball should go up and come down to be caught by the passenger in the rolling chair. This works best if the ball is thrown straight up in the vertical direction. Similarly, the coins will hit the ground at the same time. It is easier to compare the sound of the coins striking the floor than to make that comparison visually. When using a commercial apparatus, a greater distance from the floor gives a more definitive result.

Why It Works

The force of gravity and its associated acceleration is entirely in the vertical direction. Gravity does not in any way influence the horizontal velocity.

Other Things to Try

1. Place a coin at the edge of a table.

2. Slide a similar coin toward the first one, so the moving coin just knocks the stationary coin off the table and both fall to the floor. This will occur if the moving coin strikes the stationary coin at a large enough angle.

3. If a proper angle is chosen, the stationary coin is nudged off the table and falls nearly straight down. The moving coin will fall at a greater distance than the stationary coin.

The Point

Horizontal motion and vertical motion are completely independent. Excluding the effects of air resistance, the horizontally fired bullet will fall to the ground at exactly the same time as the dropped bullet. This forms the basis for an understanding of projectile motion that is greatly simplified by treating the vertical and horizontal motion separately, as if the other did not even exist.

Project 8
Target practice. Horizontal projectile—rolling off a table.

The Idea

In this experiment, you will try to hit a target. But, to improve your odds, you can use the laws of physics to predict where a projectile will land. Your projectile will be a steel ball or a marble. The physical situation is very much simplified when the projectile is shot (or launched) in the horizontal direction only. In this project, we see how close you can get to the target using the laws of physics that describe how horizontal objects move under the force of gravity.

What You Need

- steel ball or marble (to serve as a projectile)
- inclined track to get the marble rolling (This can be a piece of grooved wooden molding or a ruler with a groove)
- flat, smooth, horizontal table
- stopwatch or other timer (wrist watch, cell phone)
- cup (your target)
- meterstick
- optional: motion sensor (to measure velocity)

Method

Part 1: Find the velocity of the marble coming off the ramp

You will need this information to make your predictions.

1. Set up the ramp in such a way that its position remains fixed.

2. Place a marble at the top (or another arbitrarily mark) of the ramp.

3. Release the marble from the mark and measure the time it takes to go to the bottom of the ramp.

4. Repeat a few times until you get a consistent reading. Then, take the average. (If the ramp is too short or the slope is too great, it is more difficult to measure the time to go down the ramp.)

5. Find the final velocity at the bottom of the ramp using the equation:

$$v_f = \frac{2d}{t}$$

Part 2

1. As we found in Project 5, the vertical motion is independent of the horizontal, so we can determine the time it takes the marble to fall from the table just from the height, h, of the table. This is given by the equation:

$$t = \sqrt{\frac{2h}{g}}$$

2. Now make your prediction for how far the marble will go using $R = vt$. The distance the ball will go is now given by $R = vt$. Use the v you figured in Step 5 above and t from the previous step.

3. Set the (center of the) cup at the distance you predicted and try it out. No cheating. It is more fun to call your shot first, and then see if it works. Line the cup up visually, so it is

on a straight line with the motion of the marble, as shown in Figure 8-1.

Expected Results

Clearly the expected result is for you to have the marble roll into the cup. If the marble hits at about the distance of the cup, but to the left or right, that should count as a hit. Hitting the target requires accurate measurement of the marble's velocity on the table. It is reasonable to assume that the marble does not have any significant velocity loss for the short time it is rolling on the table.

(A simpler way of doing this—appropriate for younger readers—is to *qualitatively* compare the distance the marble goes with the height of the ramp and skipping the math. The higher the ramp, the faster the marble and the farther it goes.)

The time it takes to fall from a given distance is provided by the equation:

$$t = \sqrt{\frac{2h}{g}}$$

To use this equation, the distance the projectile falls must be compatible with the units for gravitational acceleration, *g*. If you use $9.8\,\text{m/s}^2$ for *g*, *h* must be in meters. The time to fall a given distance is shown in the following Table 8-1:

Table 8-1

Inches above floor	Meters above floor	Seconds to fall
50	1.3	0.51
45	1.1	0.48
40	1.0	0.46
35	0.9	0.43
30	0.8	0.39
25	0.6	0.36

Using this table, the distance the projectile goes is simply its velocity multiplied by the time it is in the air (from the table or equation).

Why It Works

The horizontal velocity of the marble is constant and unaffected by the fact that the marble is falling. The distance it moves is simply the horizontal velocity multiplied by the time.

The time it takes to fall a given distance is dependent *only* on the vertical *distance*.

Find the velocity at the bottom of a ramp using the fact that the final velocity is twice the average velocity divided by the time.

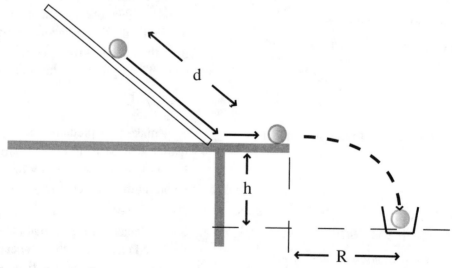

Figure 8-1 *Horizontal projectile.*

The horizontal distance the marble goes is simply the horizontal velocity times the time.

Other Things to Try

Another way to do this is to use a horizontal projectile launcher and calibrate the velocity.

The Point

Horizontal motion and vertical motion are completely independent. This means when an object is moving with only an initial horizontal velocity, the time it is in the air can be determined by how long it takes to fall.

Project 9
Taking aim. Shooting a projectile at a target.

The Idea

In this experiment, you get to shoot things around the room. You can use a toy bow-and-arrow, a toy ping-pong ball shooter, a Nerf gun, a marble launcher, or a precision projectile launcher made for this purpose. You learn to make predictions that accurately guide the projectile to the target. In this case, using the laws of physics is *not* cheating. It does, however, give you a definite advantage compared with someone who is not armed with this knowledge.

First, you measure what is the best angle to aim something for it to travel the greatest distance.

Then, you make and test predictions. To hit a target, you need to know only two things: the *velocity* of the projectile and the *angle at which* it is shot. That's all. Knowing only those two conditions, you can determine how far the projectile will go, and how high it will go. The angle is easy to measure directly, so we will first work on a simple way to determine the velocity.

What You Need

- projectile and launcher
  - A projectile launcher, such as shown in Figure 9-1. Plastic rather than steel balls are safer.
  - Or, a toy gun, a toy bow-and-arrow, a ping-pong ball shooter, Nerf gun, or a marble launcher.
- tape measure

- protractor
- target—horizontal: pan or cup; vertical: ring on a ring stand
- stool(s) or other moveable object to hold the target at the height of the launcher

Method

What is the best angle?

We start here because this part does not involve any number crunching.

1. You will be shooting your projectile from ground-to-ground or from table top to raised surface at the same height as the table top. The projectile should be launched and land at the same height.

2. Select a setting for your launcher that will fire a projectile from a given height and return to

Figure 9-1 *Projectile launcher. Courtesy PASCO.*

that same height without hitting the ceiling, a wall, or breaking anything.

3. For every test in this part, you will be using the same velocity. Pick an angle to shoot the projectile at. Launch the projectile and measure the distance. Increase or decrease the launch angle until you find the angle that gives the greatest distance. (Remember, for this part, we are measuring the distance the object goes after returning to the same height from which it was launched.)

Determine the velocity of the launcher (to make predictions).

For this part, we are going to use the method of the previous section to determine how fast the projectile is moving as it leaves the launcher. *For this part only*, we shoot the projectile *horizontally*, so we can find this velocity.

1. Fire horizontally several times and record the distance, R, that the projectile travels (in m). Take the average.

2. Measure the height when the projectile leaves the table.

3. As we did in the previous experiment, we will use the trick of *finding the time the projectile is in flight* by determining how long it takes to fall. This can be simply found just knowing the height (in meters) and using the equation, $t = (2d/g)^{1/2}$, where g is 9.8 m/s^2. Table 8-1 in the previous section gives the time, t, for various heights.

4. Now, it is a simple matter to *find the velocity* using the technique of the previous section. Divide the distance the object goes along the floor, R (in meters), by the time it was in flight, t (seconds). This is given by the formula:

$$v = R / t$$

Hitting the target

1. Select an angle, θ, at which you will shoot the projectile.

2. Predict the range, or how far the projectile goes along the floor, using the equation

$$R = (v^2/g)\sin2\theta$$

where v is the velocity you just found in number 4, g is 9.8m/s^2, and θ is the angle you selected.

3. Predict the height using h $= (v\sin\theta)^2/2g$, with the variables as defined in the previous equation.

4. Set a cup a distance R horizontally along the ground located at the same height as the launcher.

5. Set a ring on top of a ring stand at a height, h, above the level of the launcher. The circular opening of the ring should be facing the launcher. Use a few stools (stacked on top of each other, if necessary) to set the ring stand to establish the height target.

6. Visually align the targets, so they are in line with the projectile.

7. After you set the targets to where you predicted they should be, fire the launcher and see how close you get.

Expected Results

Figure 9-2 shows typical results for a projectile fired at a velocity of 10 meters per second. Notice that the 45-degree angle results in the longest range. Notice also that the 60-degree and 30-degree angles wind up in the same place. The

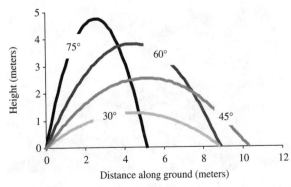

Figure 9-2 *Projectile shot at 10 m/s, returning to the same height it was shot from.*

projectile fired at 75 degrees stays in the air longer, but it has a lower horizontal velocity than the one fired at 30 degrees.

Why It Works

According to the range equation:

$$R = \frac{v^2 \sin 2\theta}{g}$$

a 45-degree angle gives the greatest distance an object moves horizontally along the ground. For a given launch velocity and a chosen angle, the range a projectile will go can be determined.

Similarly, the height equation

$$h = \frac{(v\sin\theta)^2}{2g}$$

determines the maximum height of a projectile, given the launch velocity and the chosen angle.

Other Things to Try

Combining projectile motion with "thermodynamics:" OK. The justification for doing this, other than for fun, is a stretch. But it does add a bit of extra excitement to this experiment. To do this, first of all, find a very safe place away from ceilings, loose paper, or any flammable object. Nothing flammable should be

Figure 9-3 *T. Dragoiu shows the "ring of fire" top-of-trajectory target for a projectile shot at an angle.*

underneath the ring in case of drips. Wrap the ring with a small amount of tissue paper and soak it in a little alcohol. By carefully igniting the ring, you can shoot the projectile through a flaming ring. Careful means: wear safety glasses, use a long wooden match, and make sure that neither you nor any viewers come in contact with the flame or the ring immediately after it burns because it can remain hot for a short while. This can be made even more dramatic in a very corny way by playing a recording of Johnny Cash's "Ring of Fire." This must be done in a safe place and under the supervision of an adult (if you are not yet an adult). By the way, this experiment does work perfectly well with a nonflaming ring.

Another much simpler but less accurate way to launch a projectile with a known velocity at a predictable angle is to drop a bouncy ball from a consistent height from an incline. The ball will come off at various angles, depending on the slope of the board. As a result of conservation of mechanical energy, if released from the same height above the board, the ball will bounce off at the same velocity. This may not go as far, but it provides a lower cost option to produce a reasonably constant velocity at various angles.

The Point

The range and height of a projectile can be determined from knowing *only the following two things*: *velocity* of the projectile and the *angle* that it is launched from.

Project 10
Monday night football. Tracking the trajectory.

The Idea

This experiment will take you outside to make these measurements. You can also collect data from Monday night football.

The measurement part is very simple. All you need to measure is the *total time a ball or other projectile is in the air* and the *total distance along the ground* that the projectile travels. If we measure only those two things we can figure just about everything else: launch angle, velocity, and height.

How high did the punt go? How hard was the ball hit? What angle did it go? You kick a soccer ball, hit a golf ball, and punt a football. Which has the greater velocity? Without resorting to a high-tech solution, such as a radar gun, there is a simple way to answer that question using only the laws of motion.

To do this, you either work the calculations or use the tables as a guide—your choice.

What You Need

- stopwatch
- football field
- TV tuned to a football or baseball game
- assorted projectiles and launchers: soccer ball or football; tennis ball and racket; golf club and ball

Method

Projectile

1. Launch the projectile and, at the exact same time, start the stopwatch.

2. Record how far the object goes and how long it was in the air.

Calculation

If this section contains more math than you care to do, fast forward directly to the tables in the next section.

Figure 10-1 shows a punted football from the eyes of a physicist.

1. Find the horizontal velocity (in m/s), v_x, by dividing the overall distance, R, by the total time (hang time, t).

$$v_x = R/t$$

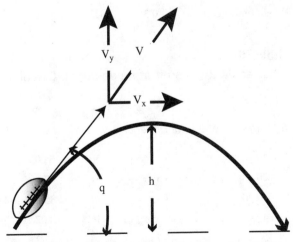

Figure 10-1 *Distance, height, velocity, and angle for a football.*

2. Find the vertical velocity (in m/s) by multiplying one half of the hang time (or the time to reach the peak) by the gravitational constant:

$$v_y = \frac{gt}{2}$$

where g is 9.8m/s^2.

3. Find the velocity (in m/s) using:

$$v = \sqrt{v_x^2 + v_y^2}$$

4. Find the angle using:

$$\theta = \tan^{-1}\left(\frac{v_y}{v_x}\right)$$

(In case you don't know what tan^{-1} is you can just use the key on your calculator with that identification. The function, tan^{-1}, also called the arctan, gives the angle if you have the *tangent* of that *angle*. You can get the tangent by dividing v_y by v_x.)

Find (or look up) the velocity, height reached, and angle launched.

Expected Results

See Tables 10-1 to 10-3.

Table 10-1
How fast it goes (in m/s)

Total Distance	Total Distance	Hang time								
		0.5 s	1.0 s	1.5 s	2.0 s	2.5 s	3.0 s	3.5 s	4.0 s	4.5 s
10 yds	9.1 m	18.5	10.4	9.5	10.8	12.8	15.0	17.3	19.7	22.1
20 yds	18.3 m	36.7	18.9	14.2	13.4	14.3	15.9	17.9	20.1	22.4
30 yds	27.4 m	54.9	27.9	19.7	16.9	16.4	17.3	18.9	20.8	22.9
40 yds	36.6 m	73.2	36.9	25.5	20.7	19.1	19.1	20.1	21.6	23.5
50 yds	45.7 m	91.5	46.0	31.4	24.9	22.0	21.2	21.6	22.7	24.3
60 yds	54.9 m	109.8	55.1	37.3	29.1	25.1	23.5	23.2	23.9	25.2

Table 10-2
How high it gets (in m)

Hang time	0.5 s	1.0 s	1.5 s	2.0 s	2.5 s	3.0 s	3.5 s	4.0 s	4.5 s
Height (in meters)	0.3	1.2	2.8	4.9	7.7	11.0	15.0	19.6	24.8

Table 10-3
What angle it goes off at (in degrees). Calculations are based on $\theta = \tan^{-1}(v_y/v_x)$

Total Distance	Total Distance	Hang time								
		0.5 s	1.0 s	1.5 s	2.0 s	2.5 s	3.0 s	3.5 s	4.0 s	4.5 s
10 yds	9.1 m	50.3	65.0	73.4	78.3	81.3	83.3	84.7	50.3	65.0
20 yds	18.3 m	31.1	47.0	59.2	67.5	73.1	76.9	79.6	31.1	47.0
30 yds	27.4 m	21.9	35.5	48.1	58.1	65.4	70.7	74.5	21.9	35.5
40 yds	36.6 m	16.8	28.2	39.9	50.3	58.6	65.0	69.8	16.8	28.2
50 yds	45.7 m	13.6	23.2	33.8	44.0	52.7	59.8	65.3	13.6	23.2
60 yds	54.9 m	11.4	19.7	29.2	38.8	47.6	55.0	61.1	11.4	19.7

Why It Works

This works for the same reasons as the previous experiment. Because horizontal and vertical motion are independent, the range and time in the air can uniquely be determined by the velocity, height, and launch angle.

Other Things to Try

Determine the velocity, maximum height, and angle for the following cases:

	Hang time	Distance traveled
Football kickoff	1.0 sec	50 yards (45.7 m)
Football punt	4.0 sec	30 yards (27.4 m)
Golf chip shot to green	2.0 sec	60 yards (54.9 m)
Pop fly behind third base	3.5 sec	50 yards (36.6 m)

The results are shown in the following table:

	Velocity	Maximum height	Angle
Football kickoff	46.0 m/s	1.2 m	23.2°
Football punt	20.8 m/s	19.6 m	21.9°
Golf chip shot to green	24.9 m/s	4.9 m	44.0°
Pop fly behind third base	20.1 m/s	15.0 m	69.8°

The Point

Knowing only the time a projectile is in the air and the distance along the ground that it travels, it is possible to determine the velocity, maximum height, and angle of the projectile.

Project 11
Monkey and coconut.

The Idea

A monkey is hanging from a branch in a tree. The monkey looks hungry and you want to throw a coconut to him. However, the monkey is nervous and, as soon as he sees something being thrown at him, he lets go of the branch. (The monkey apparently knows that in previous versions of this problem, a hunter was trying to shoot it, so the monkey is understandably a bit nervous.) Knowing the monkey will let go as soon as the coconut is thrown, where should you aim?
a) Above the monkey b) At the monkey c) Below the monkey.

What You Need

- "monkey"—(represented by a pie pan or lid of a metal container). See Figure 11-1.

- "coconut"—(represented by a projectile from Project 8)

- DC power supply

- electromagnet

- insulated wire—about 25 feet

- switch that opens the circuit at the precise moment the projectile is launched. This can be accomplished by assembling two pieces of metal foil in front of the launcher. At the instant the projectile emerges, it pushes the foil apart, opening the circuit. See Figure 11-2 for a simple way to set this up. There are also optical techniques to do this, some of which are commercially available.

Figure 11-1 *The "monkey": a metal lid held by an electromagnet attached to a vertical piece of wood.*

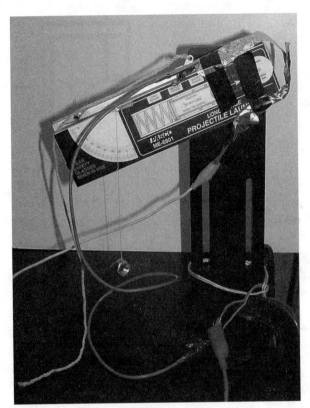

Figure 11-2 *The "coconut" is shot by a PASCO launcher with an aluminum-foil switch taped in front.*

- laser pointer
- various types of the monkey and coconut apparatus are also available commercially

Method

1. Set up the apparatus, as shown in Figure 11-3.

2. Apply the DC voltage to the electromagnet circuit.

3. Arm the launcher (by pushing the ball in with the plunger in the case of the PASCO launcher).

4. Close the switch.

5. Aim the launcher directly at the target, either visually or aided by the laser pointer.

6. Shoot the projectile. As the projectile emerges from the launcher, it causes the switch to open and deactivates the electromagnet. This releases the metal lid (monkey) as the projectile is shot at it.

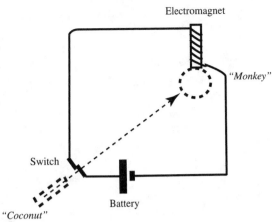

Figure 11-3 *Electrical connections for the monkey-and-coconut apparatus.*

Expected Results

The answer to the question posed previously is: b) firing at the monkey. As long as the monkey is in range, firing directly at the monkey will cause a direct hit every time.

Why It Works

Both the monkey and the coconut are subjected to the same gravitational acceleration. If the coconut is aimed directly at the monkey in the tree, the coconut will fall from that straight line path and follow the curved (parabolic) path that projectiles normally take. See Figure 11-4. As a result, the coconut falls away from that straight-line path at precisely the same rate as the monkey falls downward. This causes the monkey and the coconut to be in the same place before the monkey hits the ground.

Other Things to Try

Suppose our monkey gets tired of having coconuts thrown at him. Where should he aim a coconut of his own that he throws to deflect the one that is thrown at him? This can be set up using two projectile launchers, but it requires much more precision because the balls have a small diameter. The answer is the same as the

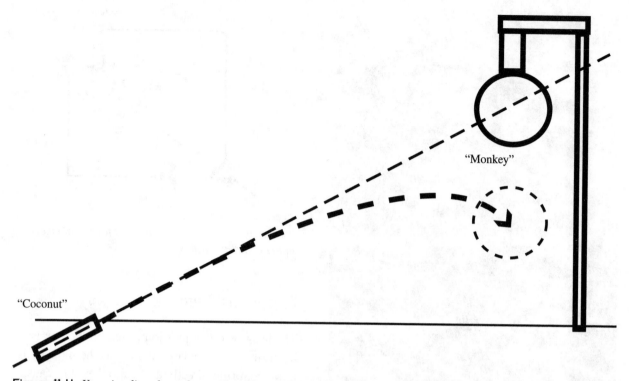

"Monkey"

"Coconut"

Figure 11-4 *You aim directly at the monkey. He will fall as fast as the coconut and will catch it on the way down.*

previous one. The monkey should aim directly at the hunter.

The Point

This project illustrates one of the underlying concepts of projectiles, which is the idea that the horizontal and vertical components of motion are completely independent and do not influence each other. The monkey does not have horizontal motion, but the coconut does. They both have vertical motion, which experiences the same rate of acceleration, regardless of the horizontal motion.

Section 2

Going Around in Circles

Project 12
What is the direction of a satellite's velocity?

The Idea

What is the direction of an object moving in a
circle? A common misconception is that the
velocity, at any given time, is pointed in a circle.
This simple experiment illustrates that the
direction of an object moving in a circular path is
in a straight line, as shown in Figure 12-1.

What You Need

- 1 marble
- roll of masking tape

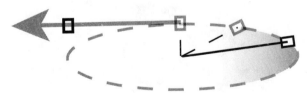

Figure 12-1 *The velocity of an object in circular
motion at any given time is a straight line tangent to
the circle.*

Method

Marble and tape ring

1. Place the marble in the center of a tape roll.
2. Get the marble spinning rapidly in a circular
   path, as shown in Figure 12-2.
3. Quickly lift the masking tape roll and observe
   the path the marble takes when it's no longer
   constrained by the tape, as Figure 12-3
   shows.

Expected Results

The marbles travel in a straight path as soon as
they are released from the roll of tape.

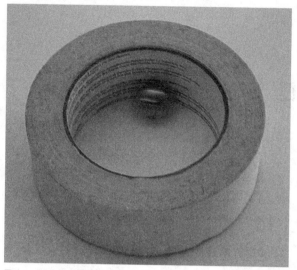

Figure 12-2 *Marble kept in a circular path by centripetal force from a tape roll.*

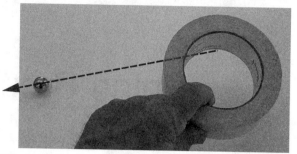

Figure 12-3 *Free marble moving along a straight path.*

Why It Works

Circular motion is the result of a centripetal force that changes the direction of motion from a straight line path to a circular path. The centripetal force is provided by the string in the case of the ball, by the interior wall of the masking tape in the case of the rotating marble, and by gravity in the case of satellites and planets. Objects traveling in a circle at any given time have an instantaneous velocity that heads in a perfectly straight line. This is actually a consequence of Newton's first law of motion, which we explore later: an object in motion tends to stay in motion in a straight line, unless it's acted upon by an external force.

Other Things to Try

Attach a string to a ball and spin the ball in a circle. Cut the string or let the string go and observe the path of the ball after it is released. Without the centripetal force provided by the string, the ball moves in a straight line.

The Point

An object moving in a circle has a velocity that takes it in a straight line at any given point in time. A centripetal force that continuously changes the direction is needed to form the circular path.

Project 13
Centripetal force. What is the string that keeps the planets in orbit?

The Idea

In this experiment, you investigate how objects move in a circle. Gravitational force keeps the planets and satellites in their orbits. The same physical laws determine how a rubber stopper on a string moves in a circle.

What You Need

- 1.5 meter of light, strong string
- 1 rubber stopper (1 or 2 holes)
- glass, plastic, or smooth cardboard tube—about 5 inches in length with a small diameter, but large enough for the string to move through freely
- spring scale—10 N
- clamp to attach the spring balance to the table
- hooked masses: 10, 20, 50, 100 g
- meterstick
- marker pen
- safety goggles—(you will be swirling an object in a circle, so safety goggles should be worn to prevent the possibility of eye injury)

Method

Set up the apparatus as shown in Figure 13-1.

1. Tie the string securely to the rubber stopper.
2. Feed the string through the glass or cardboard tube.

3. With about 1 meter of string length between the tube and the rubber stopper, cut the string, so about 25 centimeters of string is below the tube.

Making measurements

Each of these experiments uses the same basic technique. Getting the hang of it may take a little practice.

1. Put on your safety glasses. (The spinning washer poses a potential eye hazard.)
2. You have two ways to measure the centripetal force required to keep the washer moving in a circle under a given set of conditions.

Figure 13-1 *Apparatus for exploring centripetal force. Courtesy PASCO.*

39

– One way is to hang a known weight from the string, Figure 13-1 shows this approach. The force is the weight (in newtons) which is determined by multiplying the mass (in kg) by gravitational acceleration (9.8 m/s^2). This technique is simple enough, but it requires a certain degree of skill to keep the radius fixed for a given measurement.

– The other approach is to measure the force directly using a spring scale, as indicated in Figure 13-2. In this case, you need to coordinate your movements, so the force stays nearly constant for a given measurement. (Note: Holding the string at an angle slightly off vertical can introduce just enough friction to stabilize the reading while introducing an error of only a few percent.)

3. Holding the tube in one hand, swing the rubber stopper in a smooth, horizontal circle.

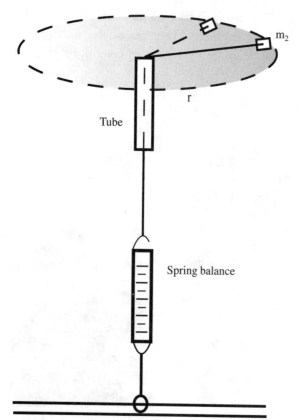

Figure 13-2 *Using a spring scale to measure centripetal acceleration.*

4. Measure how many seconds it takes to make ten rotations, and then divide by ten to get the period for one rotation. Be careful to count the first rotation at the *end, rather than at the beginning*, of the rotation. It may help to count "zero" when you start, and then to count "one" when the first rotation is completed.

5. Using the marker, place a series of marks at 1 centimeter intervals, starting at the loop for the hanging mass.

6. Using the meter stick, identify the distance between the top of the tube and the rubber stopper associated with the mark closest to the hanging weight. You can now easily measure the radius by subtracting 1 centimeter for every mark below the tube that you can count. (You can also determine the radius by measuring the length of string below the tube and subtracting from the total length of the string.) You can also use a piece of tape or a paper clip to mark the position of the string to give a radius that you measure before spinning. However you do it, make sure that nothing restricts the free movement of the string through the tube.

First investigation: Force versus velocity (for fixed radius and fixed rotating mass)

1. Set the spring balance to zero. (It's preferable that the spring balance reads directly in newtons. If it reads in grams, multiply by 0.0098 to convert to newtons.)

2. Attach the bottom of the spring balance to a clamp on the table and the other end to the string coming from the tube. See the previous Figure 13-2.

3. Start the rubber stopper going in a circle.

4. Measure the radius from the center of the circle to the rubber stopper (in meters). This should remain nearly the same for all these measurements.

5. Measure the period or the number of seconds it takes to go ten complete rotations.

6. Calculate the velocity (in meters per second) by using $v = 2\pi r/T$, where r is the radius (in meters) and T is the period (in seconds).

7. Measure the force on the spring scale while the washer is spinning. If you are using a mass hanging from the string, the force (in newtons) is equal to the weight of the mass (mass in kg times 9.8 or mass in g times 0.098).

8. Increase or decrease the velocity while maintaining a fixed radius. For each new velocity, measure the force on the spring scale. Repeat for several velocity and force measurements at (nearly) the same radius, and then plot the results.

Radius for all trials = _____ meters

	Trial 1	Trial 2	Trial 3	Trial 4	Trial 5	Trial 6
Force (in newtons)						
Time to go 10 cycles						
T = Time to go one cycle						
Velocity = 2π						

Second investigation: Force versus the rotating object's mass (for fixed radius and fixed period)

1. With the spring balance set to zero and attached to the table as done previously, start the rubber stopper spinning at a medium-paced period.

2. Measure the force and record the mass of the rubber stopper.

3. Tie a second stopper (to double the mass) at the end of the string.

4. Repeat by adding a third and then a fourth rubber stopper.

5. Complete the data table, plot your results, and describe the relationship between force and mass for fixed radius and period from your data.

Radius for all trials = (constant) _____ meters

Period for all trials = (constant) _____ seconds

	Trial 1	Trial 2	Trial 3	Trial 4	Trial 5	Trial 6
Mass (in grams)						
Force (in newtons)						

Third investigation: Force versus orbital radius (for fixed period and fixed rotating mass)

This part is more complicated than the previous two investigations and will require a greater degree of skill and patience.

1. Zero the spring balance and clamp to the table, as done previously.

2. Start the rubber stopper going in a circle.

3. Measure the radius, measure the force on the spring scale, and then measure the period as previously described. (Throughout this part of the experiment, the velocity needs to stay as constant as possible, so the only variables being studied are force and radius. Measure the period and from that determine the velocity. As the radius gets larger, it will be necessary to allow the period to decrease to maintain a constant velocity. If the velocity is reasonably close to the first reading, record the radius and the force, as well as the spring scale. Otherwise, adjust the rate of turning and try again until the velocity is reasonably close.)

4. Adjust the radius (either longer or shorter) while continuing to turn at the same rate. For each new radius, measure the force on the spring scale.

5. Repeat for several radius and force measurements at (nearly) the same period.

6. Complete the following data table and plot the results.

Period for all trials = ____ seconds

	Trial 1	Trial 2	Trial 3	Trial 4	Trial 5	Trial 6
Radius (in meters)						
Period (in seconds)						
Velocity (in m/s)						
Force (in newtons)						

Expected Results

This project leads to the following conclusions:

1. The faster the rotation (or the shorter the period of rotation), the greater the centripetal force needed to maintain circular motion.

   For a 12-gram rubber stopper, the expected results are shown in Figure 13-3. This shows the relationship is *not* linear, but that it increases more rapidly as the velocity increases.

2. The greater the mass, the greater the force needed to keep the rubber stopper going at a given speed at a particular radius. This result is expected to be linear.

3. For a given rotational speed, the shorter the string, the greater the force needed.

   For a 12-gram rubber stopper, the expected results are shown in Figure 13-4, which shows an inverse relationship between force and string length.

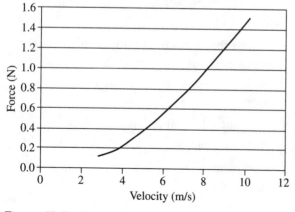

Figure 13-3 *Centripetal force versus velocity.*

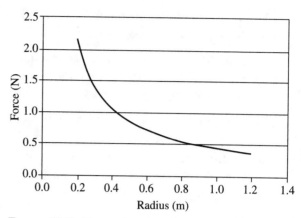

Figure 13-4 *Force versus string length.*

Why It Works

The "string" that keeps an object going around in a circle is provided by a centripetal force. In this case, it is literally a string. In the case of a satellite or planet, the "string " is the gravitational force.

The faster the object goes (for a given radius), the greater the force, according to the equation:

$$F_c = \frac{mv^2}{r}$$

where F_c is the centripetal force, m is the mass of the spinning object (the washer in our case), v is the velocity of the washer, and r is the radius of the circle.

Other Things to Try

Finding the mathematical model

Given the data shown in Figure 13-3, we can determine that force increases with the square of the rubber stopper's velocity in one of two ways:

1. Use a curve-fitting program, such as Excel. From a scatter plot, with the data selected, go to the Chart menu, select Add Trendline, and then select a power fit option. Select Add Equation to the Chart from the Options tab. This displays the mathematical model for your data. The expected result is for this to be the form $y = x^2$ or close to it.

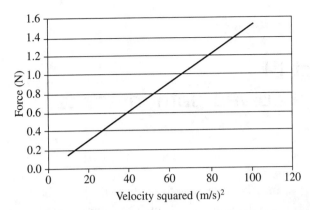

Figure 13-5 *Centripetal force versus velocity squared.*

2. Either using Excel or plotting by hand makes a graph of *force* versus *velocity squared*. If the relationship is of the form expected, that graph should be a straight line. This is shown in Figure 13-5.

Given the data previously shown in Figure 13-4, we can determine that force varies inversely with the radius (string length) using the same techniques.

1. Have Excel determine the trendline for the expected data, as shown on the graph for the previous Figure 13-4.

2. Plotting force versus the reciprocal of radius (1/r) results in a straight line, as shown in Figure 13-6.

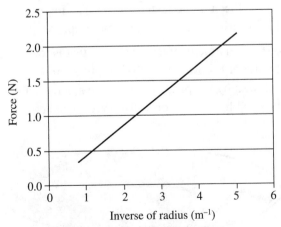

Figure 13-6 *Centripetal force versus reciprocal of radius.*

Sources of error

This project works reasonably well and enables you to find the model for centripetal force using very simple equipment. The following are potential sources of errors that may impact your results:

1. Friction between the sting and the tube overstates the required force.

2. Air resistance results in a slightly slower value of velocity.

3. At slower speeds, the circle may not be perfectly horizontal and may have a complicating effect from gravity.

Determining the accuracy of the model you found

For any of the points you measured, compare the force you measured (by either the spring scale or the hanging mass) with the expected value for the centripetal force given by:

$$F_c = \frac{mv^2}{r}$$

The Point

Centripetal force keeps an object rotating in a circle. The centripetal force equals the mass of the object times the velocity squared divided by the radius.

Project 14
A gravity well. Following a curved path in space.

The Idea

In this project, you build a simple model of a planet going around the sun. This model exhibits many of the physical properties found throughout the solar system. You can discover for yourself the basic principles of planetary motion as did Copernicus and Kepler, except you won't have to spend years squinting through a telescope on cold winter nights in the middle of the night to do this. This model also provides an intuitive way to visualize Einstein's theory that gravity is the result of a mass curving space.

What You Need

- bucket or other circular frame
- sheet of Latex, large enough to cover the opening of the bucket
- mass (roughly 50 g)—a 1-inch diameter spherical steel ball would be ideal because it can position itself in the center of the sheet
- marbles, small steel balls

Method

1. Stretch the Latex on the bucket. Remove the wrinkles.

2. Roll the marble across the sheet and observe the path it takes.

3. Now, place the mass in the center of the sheet. This should cause the sheet to become noticeably distorted. If this is not the case, it may be necessary to increase the mass, but avoid tearing the sheet. The central mass should maintain a fixed position, which can be facilitated, if necessary, by a little tape.

4. Roll a marble in a circular path around the central mass.

5. Observe the motion of the marble. See Figure 14-1.

6. Observe what happens if the marble is rolled faster or slower in a given path. What happens if the marble is closer or farther from the central mass?

Expected Results

The key observation is that the path followed by the marble is an ellipse. The path may appear circular, but elliptical paths are certainly possible. This is comparable to one of Kepler's observations concerning planetary motion.

Kepler also observed that the closer a planet gets to the sun, the faster it goes. The marbles in this experiment exhibit the same property.

If the marble is given a velocity that is too high, it will not follow the type of elliptical orbit followed by the planets around the sun but, rather, the open hyperbolic orbit followed by meteors.

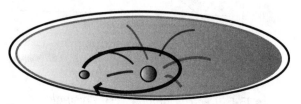

Figure 14-1

Why It Works

Kepler's law can be derived by equating the centripetal force that keeps a planet in orbit to the gravitational attraction between the planet and the sun. The depression created by the central mass exerts a force on the circulating marble that varies with position. Although this force does not exactly decrease with the inverse square of the distance, as does the gravitational attraction between a planet and the sun, it does provide a good approximation.

Other Things to Try

This experiment also provides an analogy for understanding an aspect of Einstein's theory of general relativity. The idea is that what we call gravity is really a distortion in space caused by the presence of a mass. The distortion of the sheet can be thought to represent the distortion in space, which guides the path of a planet going around the sun. As far-fetched as this may seem at first, light from stars emerging from behind the sun has been observed by astronomers to follow a bent path caused by the sun's mass, confirming Einstein's prediction.

The Point

Objects in motion around a central mass follow an elliptical path. The closer they get to the central mass, the faster they go.

Gravitational attraction can be thought of as a distortion of space caused by the presence of the mass.

Project 15
How fast can you go around a curve?
Centripetal force and friction.

The Idea

What determines how fast a car can safely go around a curve and not skid on the road? This project explores turning and friction, and how the two are related.

What You Need

- board (approximately) 36 inches by 4 inches by ¾ inch (Other shapes, including a circularly shaped board or a turntable, can also be used.)

- vertical pole, such as a ring stand, to serve as a pivot point

- a few closely matched toy cars, such as Matchbox cars

Figure 15-1 *Position of cars before rotation.*

Method

1. Drill a hole in the center of the board. The hole should be large enough to allow the board to freely rotate on the post.

2. Place each of the cars along a line running from the center to the outer edge of the board at approximately 6-inch intervals, as shown in Figure 15-1. (You can also do this with pennies or other objects instead of cars.)

3. Predict what you think will happen to the cars as you start to rotate the board around the pivot point.

4. Rotate the board, very slowly as first, but then pick up speed. What happens to the cars?

Expected Results

Cars furthest from the center begin to move first. As the cars start to move, they move away from the center, as shown in Figure 15-2.

Why It Works

The cars remain on the board as long as the frictional force is greater than the centripetal force needed to keep the cars moving in a circular path. The further you are from the center of rotation, more centripetal force is needed. For this reason, the cars furthest from the center are the first to move.

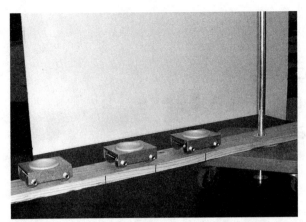

Figure 15-2 *Cars further away from the center of rotation require more friction to remain stationary.*

Other Things to Try

This can also be done using pennies on a rotating surface, such as a turntable.

The Point

Friction can provide the centripetal force needed to keep an object moving along a circular path. If the force of friction is not sufficient to provide the centripetal force for a given radius, the object will depart from its circular path.

Project 16
Ping-pong balls racing in a beaker.
Centripetal force.

The Idea

In this project, you get a pair of ping-pong balls circulating rapidly in a beaker with a blow dryer. The balls continue racing in a frantic high-speed circular path long after the blow dryer is removed. This is a fun, attention-getting demonstration that explores various aspects of circular motion, including angular velocity, centripetal force, and the effect of friction.

What You Need

- 250 mL glass beaker or plastic container roughly 5 inches (12 cm) in diameter
- 2 ping-pong balls
- blow dryer

Method

1. Place the ping-pong balls in the beaker.
2. With one hand, hold the bottom of the beaker. The other hand holds the blow dryer. (No heat is needed.)

Figure 16-1 *Photo by S. Grabowski.*

3. Direct the air from the blow dryer to rapidly circulate the air flow in a circular horizontal pattern inside the beaker.

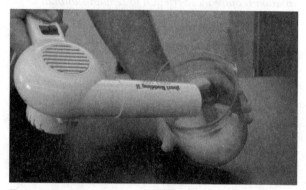

Figure 16-2 *Photo by S. Grabowski.*

4. The blow dryer should get the ping-pong balls to rapidly spin inside the glass container.

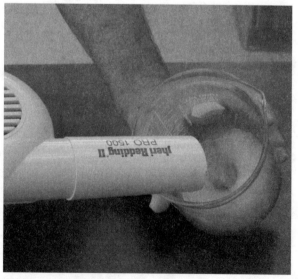

Figure 16-3 *Photo by S. Grabowski.*

5. As soon as the ping-pong balls are spinning rapidly, quickly turn the beaker upside down and (carefully) place it on the table.

Figure 16-4 *Photo by S. Grabowski.*

Figure 16-5 *Photo by S. Grabowski.*

Expected Results

The ping-pong balls continue to revolve around the inner walls of the container. While spinning, they appear to defy gravity. They also tend to move as far away from each other as they can, especially as they slow down.

Why It Works

The rapidly moving air gives the ping-pong balls kinetic energy. The inner walls of the beaker provide centripetal force that keeps the balls moving in a circular path. The force between the rotating balls and the sidewall of the beaker results in a frictional force that is large enough to hold the balls suspended above the table as they rotate. The balls have enough angular momentum to keep going until frictional forces between the ball and the walls of the container cause them to

slow down, resulting in the balls continuing to rotate more slowly and drop to the surface of the table.

The rapid rotation causes friction between the balls and the side of the beaker. This can cause the ping pong-balls to become charged, resulting in (minor) attraction to the walls of the container and repulsion from each other.

Other Things to Try

A similar effect can be achieved by vigorously rotating a pair of marbles in an inverted glass or beaker.

The Point

The ping-pong balls are given kinetic energy by the blow dryer. Like all rotating objects, their inertia tends to keep them moving in a straight line. The inside walls of the beaker apply centripetal force, which causes the path to be circular. The balls continue to move until the kinetic energy is converted into friction.

Project 17
Swinging a pail of water over your head.

The Idea

If you fill a bucket with water and turn it upside down, the water will (of course) spill out. But, if you spin the bucket over your head fast enough, you may avoid getting wet. How fast do you have to swing a pail filled with water over your head so as to not get wet? In this project, you explore what it takes not to get soaked or, in other words, how fast is fast enough?

What You Need

- small bucket with a handle or string attached
- water
- stopwatch
- meterstick
- person willing to get wet
- another person willing to get the first person wet
- optional: paper towels or a mop to wipe up spills
- optional: raincoat or umbrella

Method

1. Put some water in the bucket.
2. Predict how fast you think you need to spin the bucket to avoid spilling its contents. This can be done qualitatively by spinning at a relatively fast rotation and pushing your luck by going progressively slower. One simple refinement would be to do this in terms of

time (in seconds). A more quantitative prediction can be based on the linear or angular velocity, and it can be measured based on the person's arm length.)

3. Spin the bucket, as shown in Figure 17-1, and evaluate in terms of the predictions.

Expected Results

The slower you go, the greater the risk of soaking the spinner for a given radius. If your arm is shorter, you will have to complete the circle in less time.

Figure 17-1 *Spinning a bucket filled with water over your head.*

The maximum time to go around a vertical circle of a given radius without spilling is shown in the table below. The maximum time to spin the bucket overhead is about half that time. Keep in mind these times are based on *uniform velocity*. The most critical point of course is at the top of the circle. (If you slow down there, you may need that raincoat identified in the what-you-need list.)

Radius	Maximum time for complete circle	Maximum time to spin bucket overhead
0.5 m	1.4 sec	0.7 sec
1.0 m	2.0 sec	1.0 sec
1.5 m	2.5 sec	1.2 sec
2.0 m	2.8 sec	1.4 sec

Why It Works

The person spinning the bucket will be spared a soaking as long as the bucket moves fast enough so the centripetal force is greater than the force of gravity.

The condition for this is:

$$v = \sqrt{rg}$$

Note, this result indicates it doesn't matter how much water is in the bucket as long as the spinner moves at a sufficient speed. The larger the radius, the faster you have to go. Too much water, however, may cause the spinner to slow down.

Other Things to Try

This can also be done using confetti instead of water.

Physics alert: *there is really no such thing as a centrifugal force.* The water is given a velocity and is forced into a circular path by the centripetal force exerted by the bottom of the bucket. If the bucket is moving fast enough, the centripetal force of the bucket is needed to keep it going in a circle. If the bucket is not going fast enough, gravity would be great enough to cause the water to spill out.

The Point

The centripetal force on the water is provided by the bottom of the bucket. The handle of the bucket provides a centripetal force on the bucket itself. The water will not fall if the rate of rotation is high enough that the centripetal force is at least as great as gravity.

Section 3

Gravity

Project 18
Feather and coin.

The Idea

Does an object with a greater mass fall faster than an object with a lower mass?

This is a fundamental issue that was addressed by Galileo, as well as Apollo astronauts on the moon. After doing this experiment, you can weigh in on this question.

What You Need

- feather
- coin
- clear cylindrical plastic tube
- caps to fit the end of the tube—one closed and one with a vacuum fitting
- vacuum pump

Method

1. Put the coin and the feather in the tube.
2. Insert the end caps in each of the ends of the tube.

3. With both objects on the bottom end cap, invert the tube and let the feather and the coin fall in the tube. Make sure both are able to fall freely without interference.
4. Attach the vacuum pump to the tube and evacuate the air from inside the tube, as shown in Figure 18-1.
5. Invert the tube and observe the results again.

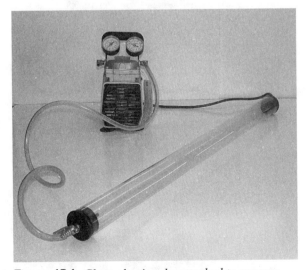

Figure 18-1 *Clear plastic tube attached to vacuum pump.*

Expected Results

With air in the tube, the coin will fall faster, as shown in Figure 18-2.

With air removed from the tube, both objects fall at the same rate, as shown in Figure 18-3.

(Two things could result in an unintended outcome, which should be avoided if possible:

Figure 18-2 *With air present in the tube, the coin fall faster.*

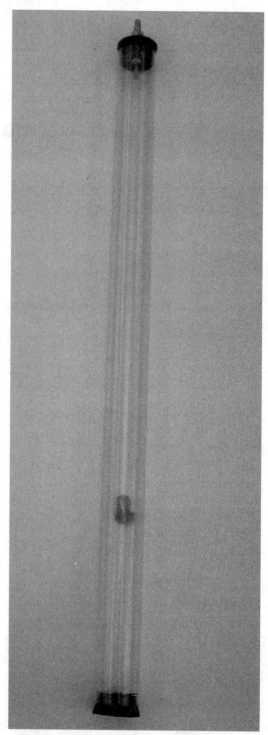

Figure 18-3 *With air removed from the tube, both objects fall at the same speed.*

With air in the tube, the coin might push the feather toward the bottom at a faster rate than it would fall on its own. Also, some electrostatic drag might develop between the feather and the plastic, which slows the descent of the feather in a vacuum.)

Why It Works

There is no doubt that the gravity of the Earth exerts a greater force on a more massive object. However, the more massive object requires precisely that same amount of larger force to cause it to accelerate. The upshot is that all objects on the surface of the Earth accelerate at the same constant rate.

Other Things to Try

This experiment has a number of variations, including:

- Compare the descent of a crumpled sheet of paper with an unfolded sheet of paper (both of the same mass).

- Compare the descent of a single pencil with several pencils bundled together.

- Tie a weight (such as a large stainless steel nut) to a string at the following intervals: 125 cm, 80 cm, 45 cm, 20 cm, 5 cm. Hold the string vertically. When dropped, each weight hits the floor in the same time interval. This is because the distance each weight falls is proportional to the *square* of the time that it is falling. These intervals are built into the spacing of the weights, so they should hit at the same time interval.

- Which falls faster (in air): a book or a dollar bill? Certainly, if they're dropped side by side, the book will fall fastest. However, if the dollar bill is placed on top of the book or below the book, the effect of air resistance will be eliminated and they will fall together.

The Point

Gravitational acceleration (in a vacuum) is a constant. Specifically, it does not depend on the mass of the falling object.

Project 19
How fast do things fall?

The Idea

Objects exposed to the force of gravity accelerate at the same rate. We proved that in the previous experiment. Here, we measure the rate of gravitational acceleration for all objects on the earth.

You measure acceleration two different ways in this experiment. In the first method, you use a stopwatch. We call this a *ballpark experiment*, which means we expect it to give a rough approximation rather than a very accurate result.

The second method involves the use of a motion sensor, which offers a greater degree of precision.

What You Need

Stopwatch method

- various objects: baseballs, golf balls, bowling balls, your physics textbook
- stopwatch
- tape measure

Motion sensor method

- motion sensor with DataStudio software
- ring stand or other support to orient the motion sensor vertically, looking downward
- basketball, softball

Method

Stopwatch method

1. Use the tape measure to identify the distance the object will be dropped.

2. One person drops the object and the other person times the trip down.

3. Start the timer just as the object is released and stop it at the precise time it hits the ground. Try to avoid anticipating the release that will give too large a time measurement and an understated value for gravitational acceleration.

4. Calculate the gravitational acceleration using the equation $g = 2d/t^2$, where d is the distance in meters and t is the time in seconds. Gravitational acceleration is measured in m/s^2, which is read as meters per second squared or meters per second per second.

Motion sensor method

1. Set up a motion detector mounted on a table with an unobstructed view of the floor, as shown in Figure 19-1.

2. Set up the motion detector to read distance versus time and velocity versus time. This can be accomplished by selecting the "velocity" file that comes with the DataStudio software package.

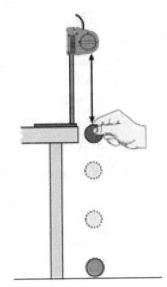

Figure 19-2 *Courtesy PASCO.*

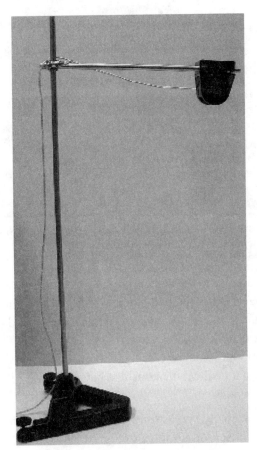

Figure 19-1 *Motion sensor aligned to measure vertical motion. Courtesy PASCO.*

3. This measurement works best by increasing the frequency of the motion sensor measurement by increasing the sampling from 10 per second to 50 per second.

4. Align the motion sensor in the vertical direction.

5. Hold the ball just under the motion sensor, as shown in Figure 19-2. Start the readings and release the ball. Try to avoid imparting any vertical momentum to the ball by letting it drop without an initial push or delayed release.

6. Capture the motion of the ball through several bounces.

7. Measure the slope of the velocity versus time graph. Use either the initial descent or the first bounce. The initial descent has the advantage of having the largest statistical

sample. The first bounce has the advantage of being free of errors associated with the release.

Expected Results

For either method, the accepted value for gravitational acceleration is about 9.81 m/s². This may vary slightly with location and elevation.

Stopwatch method

For a typical outdoor high-school athletic bleacher about 15 feet above the ground (about 4.6 meters), an object will take about 1 second to fall. We learn in the next project that a person's reaction time can easily be as much as ¼ second. As a result, any given measurement may have an error of as much as about 25 percent. (This can be even greater because there can be non-offsetting errors for the start and stop time of the measurement.) This is not very precise, but it puts us in the ballpark. It is hard to improve on this because of the limitation in measuring time inherent in the use of a stopwatch. Some people find that listening for the ball to hit the ground is easier to time than trying to observe it visually. A greater distance to fall also reduces errors because the reaction time is a smaller percentage of the overall time measured.

The following chart summarizes expected times for various distances. Times measured in this range gives reasonable values for gravitational acceleration, g.

Distance (feet)	Distance (meters)	Time to fall (seconds)
5	1.5	0.6
10	3.0	0.8
15	4.6	1.0
20	6.1	1.1
25	7.6	1.2
30	9.1	1.4
35	10.7	1.5
40	12.2	1.6
50	15.2	1.8
55	16.8	1.8
60	18.3	1.9
65	19.8	2.0

Another result expected is that, within the accuracy of this experiment, all objects fall at the same rate of acceleration, regardless of their mass.

Notice how sensitive the results are on the time measurement. For instance, suppose you drop a bowling ball from a 4.6 meter height and measure 1.1 seconds instead of 1.0 seconds. That 0.1 second error would result in a calculated value for gravitational acceleration of 7.6m/s^2 instead of the expected value of 9.8 m/s^2 or a 22 percent error. A 0.1 second error is less than the reaction time of most people so it is a good thing that we have another way to make this measurement.

Motion sensor method

With a motion sensor, the range of measurements is much tighter. The position versus time graph is shown in Figure 19-3. Notice this shows a curved line typical of acceleration. As the ball falls, the position increases, s the portion of the curve

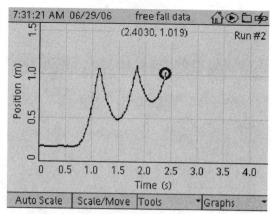

Figure 19-3 *Position versus time for a falling ball showing two full bounces. Courtesy PASCO.*

sloping up to the right represents the falling motion. After the ball bounces off the floor, the distance increases, which generates the curved line that slopes down to the right. This graph shows an initial release and then two bounces. The data collection stops just before a third bounce.

A velocity versus time graph generated by a motion sensor is shown in Figure 19-4. Gravitational acceleration is given directly by the slope of the line. This can be determined by dividing the rise (change in velocity) by the run (corresponding change in time). The slope can also be found by using the slope tool located in the DataStudio pull-down menu. This graph shows the same drop followed by two bounces, as you saw in Figure 19-3. Notice the first bounce occurs just before 1.2 seconds. The ball reaches its first peak at 1.5 seconds and begins to fall again. In Figure 19-4, the velocity rapidly changes from positive (above the line) to negative (below the line).

Also notice one interesting aspect of the physics of free-fall, illustrated by Figure 19-4. After each bounce, the slope is the same *below* the zero line (bouncing up), at the zero line (at the highest point) and *above* the zero line (falling back down). What this means is gravitational acceleration is *constant* and affects an object in free-fall, *regardless of whether it is moving up or down.*

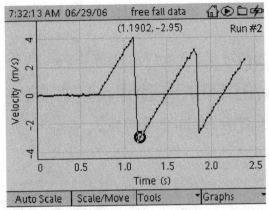

Figure 19-4 *Velocity versus time for a falling ball. The slope of each line gives the acceleration of the ball in free fall. Courtesy PASCO.*

Why It Works

Part 1 is a direct measurement and application of the basic motion formula:

$$a = 2d/t^2$$

where we find the acceleration due to the force of gravity.

Part 2 measures the same thing, but it uses a much more precise measurement of the distance traveled in a given time. We know from Projects 1 and 2 that the slope of the distance versus the time graph gives a measure of velocity. Similarly, the slope of the velocity versus the time graph gives acceleration. Each bounce provides a replication of this experiment that can provide a separate data point.

Other Things to Try

A motion sensor reveals the brief time that a ball encounters the ground as it compresses, de-compresses, and eventually reverses direction. Some balls do this more quickly than others. This can be seen in time-lapse photography but can also be noticeable in the distance versus time graphs generated by motion sensor.

There is another method for measuring the Earth's gravitational acceleration using a pendulum. See Project 22. Compare this with the results you get with the motion sensor.

The Point

This experiment gives two ways to measure the acceleration on any object caused by the gravitational force of the Earth. The first way is a direct measurement limited by the reaction time to record how long it takes an object to fall. The second method uses a motion sensor that captures this data with greater resolution and precision, and when interpreted graphically gives a more accurate value for gravitational acceleration. In either case, the correct value is 9.8 m/s^2 or 32 ft/s^2.

Project 20
The buck stops here (the falling dollar).
Using a meterstick to measure time.

The Idea

This experiment explores the nature of *free-fall*: the longer an object falls, the greater the *distance* it falls. Measuring the distance an object falls can give an indication of the *time*. This can be used to estimate a person's reaction time. You use both a dollar bill and a meterstick to prove this point.

What You Need

* meterstick

Method

1. This requires two people. The first person holds a meterstick upside down, so the end that reads 0 cm is directed downward.

2. The second person holds their fingers at the bottom of the meterstick ready to grab the meterstick, as shown in Figure 20-1.

3. The meterstick is dropped, and then caught as shown in Figure 20-2.

4. The *distance* the meterstick falls is an indication of the person's reaction *time*. Under gravitational acceleration, distance is related to time according to the equation $d = \frac{1}{2} gt^2$ where g is the gravitational acceleration constant, 9.8 m/s^2, and time is measured in seconds. This equation gives the distance in meters. This relationship is tabulated in Table 20-1:

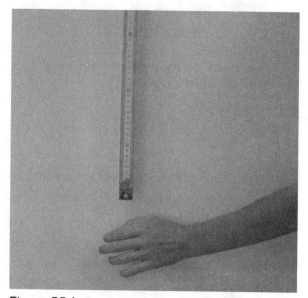

Figure 20-1 *Ready to catch the meterstick.*

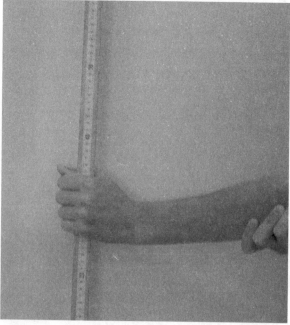

Figure 20-2 *The position where the meterstick is caught is an indication of the time it was falling.*

Table 20-1

Distance (meters)	Distance (centimeters)	Reaction Time (seconds)
0.04	4.0	0.09
0.06	6.0	0.11
0.08	8.0	0.13
0.10	10.0	0.14
0.12	12.0	0.16
0.14	14.0	0.17
0.16	16.0	0.18
0.18	18.0	0.19
0.20	20.0	0.20
0.22	22.0	0.21
0.24	24.0	0.22
0.25	25.0	0.23

Expected Results

The reaction *time* can be determined by the *distance* that the meterstick falls before being caught. The meterstick will typically fall about 10–20 centimeters before being caught, but this will vary with the individual.

Why It Works

The distance an object falls increases with the *square* of the time it falls. Similarly, the time it takes to fall is proportional to the *square* root of the distance.

Other Things to Try

A dollar bill is about 15.2 cm (6 inches) in length. According to the previous chart, it will take a dollar bill nearly 0.18 seconds to fall.

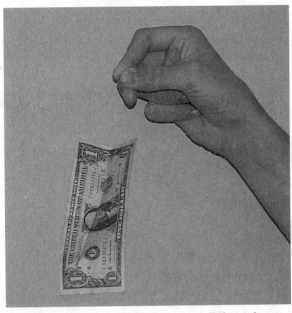

Figure 20-3 *Money often seems to fall through our hands. It falls through its own length in a time less than most people's reaction time.*

Challenge someone to catch the dollar. Unless the person anticipates that release, the bill will fall (almost every time).

Typical human reaction time is about ¼ second. Most of the time, people are unable to catch the bill. Occasionally, someone can get lucky and anticipate the falling dollar. If you are offering to let the person keep the bill if they catch it, you may want to consider a smaller denomination.

The Point

The distance an object takes to fall is related to the time it takes to fall that distance. Knowing the time lets you predict the distance. Similarly, knowing the distance lets you predict the time.

Project 21
Weightless water. Losing weight in an elevator.

The Idea

When things fall, they no longer *seem* to have weight. Objects, including people, float in the Space Shuttle as if there were no gravity. The shuttle orbit is not that high above the earth's surface for the gravitational attraction to the Earth to disappear and, if the shuttle were not moving so rapidly in orbit, it would be pulled straight down to the earth. So, where does the force of gravity go to make the shuttle astronauts seem weightless? It has to do with the forces on falling objects. In this experiment, you investigate what happens to the weight of falling objects using a falling cup of water and by holding a weight on a scale in an elevator.

What You Need

- cup with a few holes in it
- water
- spring scale
- mass
- optional: an elevator

Method

Cup

1. Fill the cup with water. Observe what happens.
2. Drop the cup (over a sink or bucket). Observe what happens.

Elevator

1. Hang the mass on the spring scale.
2. With the mass held stationary, note the weight of the object.
3. While holding the scale, let the scale and mass drop toward the floor. Stop both just before they hit the floor.
4. You don't have much time to do this, but observe the scale reading as it first starts to fall and the reading as it is slowed prior to hitting the floor.
5. Again, while holding the scale, raise the scale and mass from the floor. Bring it to a stop while you're still holding it.
6. If you can get to a real elevator, observe the scale reading with the mass suspended as the elevator goes up and as the elevator goes down. Even in a real elevator, you will find the period of acceleration for you to make these observations is short because elevators reach a steady velocity fairly quickly.

Expected Results

The water drips out from the holes when the cup is filled with water and held stationary. The water stops dripping when the cup is in free-fall.

The scale reads a greater weight when you lift the mass. With your arm fully extended as the scale and mass slows, the scale reads a lower reading than while stationary.

The scale reads a lower weight when you drop the mass. As you slow the scale and mass as it nears the floor, the scale reads a higher reading than while stationary.

With the elevator, the scale reads a higher weight when it first starts to go up and a lower weight as the elevator slows to the next stop. Going down, the scale reads lower at first, and then higher at the next stop.

Situation	At first scale reads	At end scale reads
Hand-held scale moving up	Higher	Lower
Hand-held scale moving down	Lower	Higher
Elevator moving up	Higher	Lower
Elevator moving down	Lower	Higher

If the starting-up and slowing-down phase is uniformly spread over about 1 second, the change in apparent weight measured by the scale should change (very briefly) by not quite 50 percent. These changes are illustrated in the following Figures 21-1, 21-2, and 21-3.

Figure 21-2 *Upward acceleration.*

Figure 21-3 *Downward acceleration.*

Figure 21-1 *Object at rest.*

Why It Works

With the cup held, the water is drawn through the openings by gravity. However, when the cup is in free-fall, it falls at the same rate as the water. While falling, the water is (apparently) "weightless," similar to astronauts in the shuttle. If the cup wasn't falling, the bottom of the cup would resist that pull, leaving the water no other resort than to fall out of the holes. However, with the cup falling, the water does not experience a force from the cup opposing its downward movement. Because the cup was falling, the water in the cup seemed weightless.

We have all seen images of astronauts floating in the shuttle and space station, as if they were "weightless." As a tube of scrambled eggs floats by an astronaut, it certainly appears that way. However, *weight* is the force caused by gravity on a mass and, although that attraction drops off as the inverse square root of the distance, it never becomes zero. In fact, at the shuttle orbit of 200 km, the force of the Earth's gravitational attraction is only about 6 percent lower than what it is on the surface of the earth.

Objects in orbit are essentially falling at a speed consistent with maintaining the orbit. The apparent weightlessness of objects in space is the result of the objects in the shuttle falling to Earth at the same speed as the shuttle itself is falling. Because the water is falling at the same speed as our cup, it, too, appears weightless. When we hold the cup, we balance the force of gravity on the cup only, but not on the water. In that case, the water has weight that spills out of the holes in the cup.

The weight measured on the scale is a combination of the actual weight decreased or increased by the effect of accelerating the weight. It *does not matter how fast* the weight is moving. Once a high-speed elevator gets going at its cruising velocity, the measured weight should exactly equal the stationary weight. It is only the acceleration experience during the stopping and starting that affect the force on the object during that time.

Other Things to Try

A visual accelerometer can be used to indicate the direction and relative magnitude of the accelerations that accompany the weight changes encountered here.

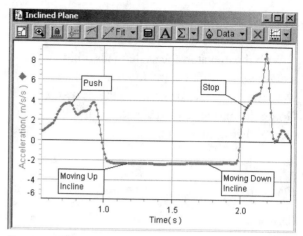

Figure 21-4 *Courtesy PASCO.*

Using a motion sensor to measure an object moving up and then down an incline provides a closer look at the acceleration of an object subjected to gravity. Figure 21-4 shows the acceleration versus time displayed in DataStudio. This shows that acceleration is downward (slightly more than -2 m/s^2) for both the uphill and downhill segments of the graph.

The Point

The sensation of weight is caused by a force (such as the ground) holding you up. This upward (or normal) force, whether exerted from below (as in the cup) or above (as with the scale and weight), is how we experience weight. If the object supporting us is also falling, we are no longer exposed to the force we experience as weight.

Project 22
What planet are we on? Using a swinging object to determine the gravitational acceleration.

The Idea

This project explores an indirect way of measuring the gravitational acceleration of the Earth (or any other planet you may do this experiment on). Because gravitational acceleration affects how fast a pendulum swings, we can take advantage of that to find the gravitational acceleration provided by measuring two things: how *long* the pendulum is and how much *time* it takes to swing back and forth. (If you are less than eighteen years old, please be sure to get your parents' permission for any interplanetary travel for this project.)

What You Need

- pendulum consisting of a mass supported by a string attached to a support

- pendulum with a long string and a large mass such as a bowling ball supported (safely and securely!) from the ceiling

- stopwatch

Method

1. Measure the period of a pendulum by measuring how long it takes for the pendulum mass to swing back and forth one time. Since this may be less than a second, more accurate measurements can be made by counting the time for 10 back-and-forth excursion, and then dividing by 10. Remember that in counting the cycles, the first cycle is counted when the mass returns to the point from which it was released and not at the point when it is first released.

2. Measure the length of the pendulum. This is the distance in meters from the center of the hanging mass to the point of attachment.

3. Try to minimize vibration of the ring stand or other support structure. Also keep the pendulum moving in two dimensions. Some people like to use a double string—one on either side of the mass—to keep the pendulum from wobbling.

4. Calculate the gravitational acceleration, g (in meters per second per second), using the equation:

$$g = \frac{4\pi^2 L}{T^2}$$

where L is string length in meters and T is time in seconds. (This will also work if you measure L in feet but you will get an answer in feet per second per second.) Try this multiple times and take the average to get the most accurate result. See Figure 22-1.

Expected Results

Gravitational acceleration should be close to the accepted value of 9.81 m/s². Results within 2 percent of this value are easily achievable. If you are working with feet, gravitational acceleration is 32 feet/s². Longer pendulum lengths encounter less frictional loss and are easier to get an accurate period measurement. Remember, 100 centimeters is equal to 1 meter when determining the length of the pendulum.

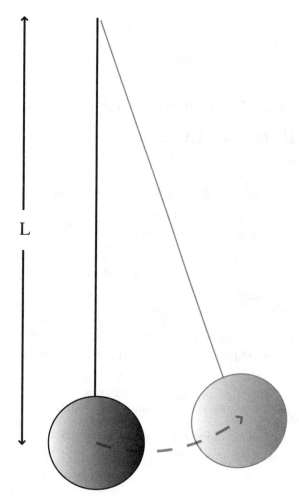

L

Figure 22-1 *Using a pendulum to find g.*

The following set of values results in the 9.81 m/s² target value for gravitational acceleration:

Pendulum length in centimeters	Pendulum length in meters	Pendulum period resulting in g = 9.81 m/s²
10	0.1	0.6
20	0.2	0.9
30	0.3	1.1
40	0.4	1.3
50	0.5	1.4
100	1.0	2.0
200	2.0	2.8
300	3.0	3.5
400	4.0	4.0

Why It Works

It stands to reason that the greater the pull of gravity, the faster the pendulum motion and the shorter the period. This is given by the equation for the period of a pendulum:

$$T = 2\pi\sqrt{\frac{L}{g}}$$

where *L* is the length (in meters) and the period is one cycle back and forth (in seconds). Solving this for gravitational acceleration gives us:

$$g = \frac{4\pi^2 L}{T^2}$$

The definition of the period of a pendulum is the number of seconds for it to swing *back and forth* one time.

Other Things to Try

You have been captured by alien abductors and taken to an unknown planet (where there are no video games and no cable TV). You are able to remove your shoe, which has a 15 cm (0.15 m) shoelace. You find that when you let your shoe swing freely in a short arc, it returns to the point from which it was released in 1.26 seconds. To which planet should you direct the interplanetary rescue team? The gravitational acceleration on the various planets is: Venus 8.93 m/s², Earth 9.81 m/s², Mars 3.73 m/s², Jupiter 924.9 m/s², and Saturn 10.6 m/s². Try it. (Hint: the answer is this planet is one of Earth's neighbors in space possessing a very thin atmosphere, ice caps, and a reddish clay-like surface.)

The Point

For a given string length, the period of the pendulum depends on the gravitational acceleration. This provides a fairly accurate method for measuring the local gravitational acceleration.

Force and Newton's Law

Project 23
Newton's first law. What to do if you spill gravy on the tablecloth at Thanksgiving dinner.

The Idea

An object at rest (including an object at rest on top of a tablecloth) tends to stay at rest unless acted on by an external force. One way to prove this is to pull a cloth out from under the object. This can be done more simply at first or more elaborately as you build your confidence in the law of physics.

What You Need

- tablecloth (one with low friction is best— tweedy fabrics, gravy spots, and spilled soda can increase friction and compromise the intended results)
- table
- objects to place on the tablecloth such as bowls, bottles, lit candles, and your physics textbook
- spare roast turkey, stuffing, and cranberry sauce, if you actually attempt to do this at *your* Thanksgiving dinner

Method

1. Place the cloth on the table, so at least several inches extend beyond the edge of the table.

2. Carefully place the objects on the cloth. If the objects are closer to the edge of the table this is easier, but allow for at least a few inches for the objects to slide.

Figure 23-1 *Don't hesitate!*

3. At this point, all you need to do is pull the tablecloth out from under the objects on the table. As with a band-aid, *the faster the better*. Don't hesitate and be tentative with your pull because that increases the chances for the objects to topple. (It doesn't hurt to create suspense by pretending you never did this before and have every reason to expect it to fail. The more you do this, the greater level of acting skills this may require.)

Figure 23-2 *A fast pull works better than a tentative pull.*

4. You can also do this with just one beaker and a cloth. Perhaps this is a less-dramatic start, but it still proves the same point.

Note: This works best when objects placed on the tablecloth have smooth bottom surfaces. Bowls with a circular lip tend to catch on the tablecloth. Pottery with a felt bottom or mounted silicon rubber resting points can also lead to humiliation and ridicule if they hang up during this demonstration. The objects should have a short vertical "moment arm," which means bowls are safer than bottles and partially filled bottles are safer than empty bottles. Bottles with liquid should be at room temperature to avoid condensation on the outside of the bottle, which can increase friction.

Expected Results

The cloth is removed and the objects-at-rest sitting on the tablecloth tend to stay at rest in approximately the same position they were originally placed. Most likely, there will be some sliding and even teetering before the objects come to rest.

The criterion for stability is that the height to diameter ratio for cylindrical objects be less than the coefficient of static friction between the cloth and the object.

Why It Works

This is basically a fun experiment, but there is a good bit of physics to learn here. The objects retain their positions on the table due to *Newton's first law*, which states that an object in motion tends to stay in motion unless acted upon by an external force. (An object at rest tends to stay at rest *unless acted upon by an external force*.) If excessive friction exists between the cloth and the bottom of the objects, there will be an external force and the objects will move. The table must have low enough friction so the tablecloth can be pulled out smoothly, but enough friction so the objects don't slide too far after the cloth is removed. The small frictional force that occurs when the cloth is pulled out exerts a torque that can rotate the object, especially one whose center of mass is relatively high above the table. The frictional force exerted by the table on the bottom surface of the objects opposes this rotational motion and helps stabilize tall objects, such as bottles and candlesticks.

Other Things to Try

As a way to get in touch with your inner nerd, you can draw diagrams, called *free-body-diagrams*, showing all the forces present in this project. You can learn a lot of physics by doing this.

The Point

One aspect of Newton's first law is that an object at rest tends to stay at rest. This can be seen in the reluctance of the objects on the table to be moved as the cloth is pulled out from under them. Some frictional force exists between the cloth and the objects, which exerts a torque that, if strong enough, will rotate and topple the object.

Project 24
Newton's first law. Poker chips, weight on a string, and a frictionless puck.

The Idea

This experiment further explores Newton's first law in both the horizontal and vertical directions.

What You Need

- 5 to 10 poker chips (or coins)
- table
- string—strong enough to support the mass, but weak enough to break when pulled
- 1 weight with attachment points on both the top and bottom
- 1 support to hang the weight

Method

Chips

1. Place the chips in a vertical stack on the table.
2. The table should be smooth enough for the chips to slide freely across its surface.
3. Take one chip and direct it toward the stack by flicking it with your fingers or pushing it rapidly toward the stack.

Weight on a string

1. You are going to do this twice (two different ways), so if you have enough materials, it works best if you duplicate the set-up side-by-side.
2. Use the string to hang the weight from the support.
3. Attach a string on the bottom of the weight.
4. Predict what will happen when you pull the string.
5. First time—pull the string slowly.
6. Second time—pull the string quickly.

Expected Results

The sliding chip should knock out the bottom chip and take its place in the stack (Figure 24-1).

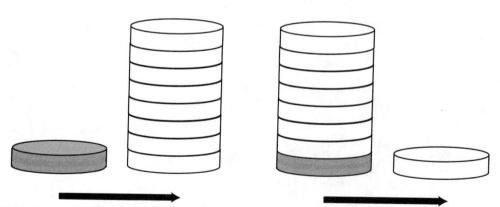

Figure 24-1 *Inertia keeps the upper chips in place while the lower one is removed.*

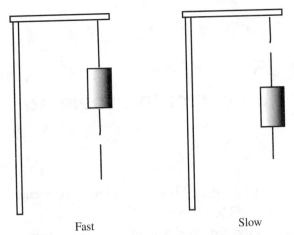

Fast   Slow

Figure 24-2 *Where the string breaks depends upon how fast you pull.*

Pulling the string slowly causes only the upper string to break.

Pulling the string quickly causes only the bottom string to break.

Why It Works

These are simple demonstrations of Newton's first law. The stack of poker chips remains are rest. The momentum of the moving chip is transferred to the chip it replaces. Momentum is explored in Section 5 in this book.

When the string is pulled slowly, the force from pulling is added to the weight pulling down on the upper string. The combined tension is greater on the upper string and that is the string that breaks.

When the bottom string is pulled rapidly, the mass, which is at rest, tends to stay at rest and the tension is applied to the bottom string, which breaks.

Other Things to Try

You can explore Newton's first law in a number of other ways. These include:

1. Cut or tear a rectangular sheet of paper nearly in thirds, leaving just a short ⅛ inch (1 mm) piece of paper remaining to hold the sections together. Challenge someone to pull sideways at both ends (perpendicular to the tears) to

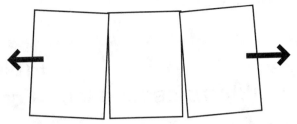

Figure 24-3 *It is just about impossible to make the center piece of paper fall by pulling the other two pieces sideways.*

cause the center section to drop. Because of Newton's first law, this is virtually impossible.

2. Place a handful of coins on your inner arm while it's bent. In one quick motion, swing your arm forward and catch the coins in midair. In the first one-tenth of a second, the coins fall only about 2 inches (or 5 centimeters), so if you are quick, you stand a good chance at catching them. This takes practice. Make sure no one gets hit, either by the coins or your arm.

3. Place a coin on a card placed directly over the bottle. Flick the card away and the coin drops into the bottle.

4. Support a coin or sugar cube on the edge of an embroidery hoop balanced on the opening of a jar (or bottle). Smoothly pulling the hoop will result in the coin or cube falling into the jar below.

5. Slide an air puck or a slider on an air track. (An air hockey table can also work.) Without friction, an object keeps moving in a straight line until a force interacts with it, just as an object in space. This demonstrates the aspect of Newton's first law that refers to a body in motion staying in motion.

The Point

This project explores Newton's first law, which is also known as the *law of inertia*: a body at rest tends to stay at rest unless acted upon by an external force. A body in motion tends to stay in motion in a straight line unless acted upon by an external force.

Project 25
Newton's second law. Forcing an object to accelerate.

The Idea

This classic experiment explores the connection between an object's acceleration and the force applied to it. This fundamental principle of physics was first formulated by Sir Isaac Newton in the famous second law of motion that bears his name. To measure acceleration, you use either the stopwatch or the motion sensor technique of measuring acceleration, which we used in previous experiments. The force will be provided courtesy of the Earth, in the form of the gravitation force on a mass hanging from a string.

What You Need

- low-friction cart (or an air track and glider, if available)
- spring scale
- mass set (including 50 g, 100 g, 200 g)
- tape
- string
- pulley (low mass and low friction is preferable)
- clamp to attach the pulley to the table
- table top (at least 1 meter in length)
- stopwatch and meterstick or motion sensor

Method

1. Determine the mass of the cart in grams. Divide by 1000 to get kilograms.

2. Place a 100g (0.1kg) mass in the cart. Secure it with tape, if necessary.

3. Set the cart at one end of the table, and attach the pulley to the other end.

4. Attach the string to the cart, run it over the pulley, and tie a loop that extends a few inches below the edge of the table, in the other end, as shown in Figure 25-1.

5. While holding the cart in position at the far end of the table, hang a mass on the loop on the other side of the string.

6. Next, you release the cart and let the weight of the hanging mass pull the cart across the table. As you do this, you measure the acceleration of the cart using either of the previous methods:

   – Stopwatch: measure the time (in seconds) for the cart to be pulled a measured distance (in meters). The acceleration (in m/s^2) is determined by a = 2d/t^2, where d is the distance that the object is pulled across the table (in m) during time, t (in seconds).

Figure 25-1 *Newton's second law apparatus. Courtesy PASCO.*

- Motion sensor: record the position of the cart as it is drawn across the table. Display the velocity versus time graph and determine the acceleration of the cart by finding the slope of that graph. This can be done either using the Slope tool from the DataStudio menu or more simply by obtaining the acceleration as the change in velocity divided by the change in time.

7. Repeat this measurement, but make the following changes:

   - Vary the mass in the cart, but keep the applied force constant, as indicated in Figures 25-2 and 25-3.

   - Vary the applied force by adding or removing some of the *hanging weight*, but keep the mass in the cart constant, as shown in Figure 25-4.

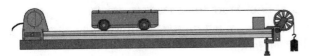

Figure 25-2 *Finding acceleration as a function of mass, while keeping the force constant. Courtesy PASCO.*

Figure 25-3 *Adding mass to the cart while keeping the force constant. Courtesy PASCO.*

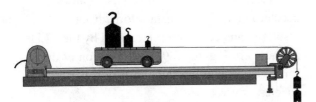

Figure 25-4 *Acceleration as a function of the hanging mass. Courtesy PASCO.*

Proving Newton's second law

Newton's second law, which states that F = ma, or as Newton originally put it, a = F/m.

- *m* represents the *entire mass* of the system and includes the *mass of the cart (m_c), plus the mass in the cart (m_1) plus the hanging mass (m_2).*

- *F* is the applied force that pulls the cart and is given by the hanging mass, m_2 (in *kilograms*, not in grams) times the gravitational acceleration (9.8 m/s²). (To get kilograms from grams, divide the number of grams by 1000.)

- *a* is the acceleration (in m/s²) of the *entire system*, including the cart, its contents, and the hanging mass.

You can use the following to organize your data:

Mass (kg) of cart m_c			
Mass (kg) in cart M_1			
Hanging mass (kg) M_2			
Total mass (kg) = $m_1 + m_2 + m_c$			
Force (in *N*) = $9.8 \times m_2$			
$a = 2d/t^2$ or from motion sensor			

Expected Results

The effect of force on acceleration

For a fixed mass in the cart (m_1), the greater the applied force, the greater the acceleration. The expected relationship between acceleration and force is shown in Figure 25-4, which (for simplicity) shows the effect of increasing the hanging mass on acceleration. (The actual driving force is given in newtons, which is simply 9.8 times the mass in kilograms.) Without friction (and to the extent that friction is eliminated from this experiment), this should be a linear relationship as indicated in Figure 25-5.

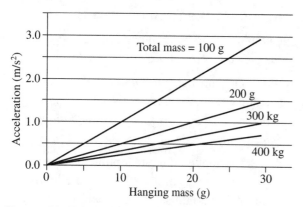

Figure 25-5 *Predicted acceleration versus applied force for different values of (total) mass.*

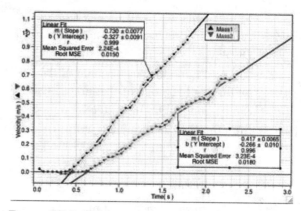

Figure 25-7 *Velocity versus time for two different masses accelerated by a constant force. The slope of the v-t curve gives the acceleration. Courtesy PASCO.*

The effect of mass on acceleration

For a given applied force, the heavier the load, the smaller the rate of acceleration. This is an *inverse* relationship, as shown in Figure 25-6.

Experimental results for acceleration for a given mass and applied force come close to the predicted results if the frictional forces are not significant. Even with friction, it can still be shown that acceleration depends on applied force and is inversely proportional to the mass. Friction increases when too much mass is placed in the cart. However, if the mass is too small, the acceleration can be so high, it becomes more difficult to measure accurately.

Use of low-friction tracks reduces the amount of friction. Motion sensors provide a nice way to determine the acceleration. Figure 25-7 shows the

result of motion sensor data for two different total accelerated masses.

Why It Works

Newton's second law states that $F = ma$ or $a = F/m$. More force leads to greater acceleration, but more mass lowers the rate of acceleration.

Other Things to Try

You may want to consider doing this using a Hover Puck drawn across the floor by a mass hung from a pulley, as shown in Figure 25-8. As before, remember to include the mass of the Hover Puck as part of the total system mass being accelerated. The higher the pulley is supported above the floor, the longer the run you can have across the floor. A qualitative but very intuitive way of showing the relationship between a force and acceleration can be shown using an LED accelerometer. The constant force from the fan results in an acceleration indicated by the LEDs as shown in Figure 25.9. The direction of the force vector is in the same direction as the acceleration vector.

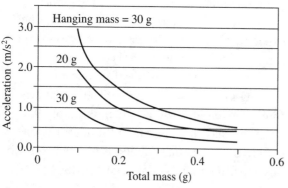

Figure 25-6 *Predicted acceleration versus total mass.*

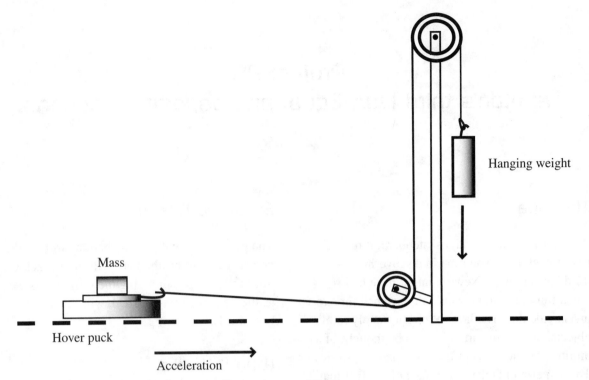

Hanging weight

Mass

Hover puck

Acceleration

Figure 25-8 *Using a Hover Puck to prove Newton's second law.*

Figure 25.9 *The fan exerts a force which accelerates the cart. The LEDs on the accelerometer show how the cart accelerates. Courtesy PASCO.*

The Point

The result here is one of the most significant results in physics: a force causes acceleration. For a given mass, the acceleration of an object is proportional to the applied force. For a given force, the acceleration is inversely proportional to the amount of mass.

Project 26
Newton's third law. Equal and opposite reactions.

The Idea

If Sir Isaac Newton had a skateboard, it might have saved him some time in discovering his third law, although Newton might have had so much fun doing it, he wouldn't have had time to invent calculus. The laws of physics apply to all objects. Sports, in particular, can be thought of as intuitive applications of the principles of physics. This experiment takes advantage of the fact that all objects in the universe follow the laws of physics. We focus particularly on Newton's third law and conservation of linear momentum.

What You Need

- 2 rolling chairs
- or 2 people capable of keeping their balance on skateboards (each with helmets); rollerblades will also work
- medicine ball or a several pound mass, such as a bowling ball
- safe place to do this

Method

1. Two people face each other sitting in the chairs on rollers, a few feet apart.
2. One person tosses the medicine ball to the other (both are seated in chairs). Feet should be kept off the floor, so the chairs are free to move.
3. The two people again face each other. One tries to push the other. What happens?

Expected Results

The person who catches the ball, as well as the person who throws the ball, will move backward. Similarly, the person doing the pushing, as well as the person getting pushed, will recoil backwards.

Why It Works

Momentum is mass times velocity.

At the start of this, the two skateboarders have zero momentum (they have mass, but no velocity, so their momentum is zero).

The velocity of the ball transfers momentum from the first to the second person. The first person recoils backward. The second person also moves backward in the opposite direction.

Another principle illustrated here is Newton's third law: For every action (movement of the ball), there is an equal and opposite reaction (recoil of the person in the chair).

Other Things to Try

The previous demonstrations can be done by skateboarders or rollerbladers. (Please remember, although we are interested in horizontal action and reaction here, gravity is still active in the vertical direction, so keep your balance.)

Mousetrap and tennis ball

Conservation of linear momentum and Newton's third law can be demonstrated by attaching a mousetrap to a low-friction cart. The trap is set and a tennis ball is positioned in place of the cheese. When the mousetrap is released, the process of tossing the ball results in the equal and opposite reaction of the mousetrap recoiling in a backward motion. This is shown in Figure 26-1. Both the mousetrap and the ball initially have zero momentum. The momentum of the ball going to the left is equal but opposite to the momentum of the mousetrap and cart moving to the right.

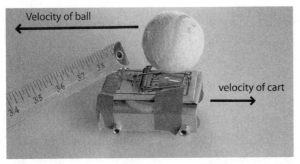

Figure 26-1 *Equal and opposite reactions.*

Fan car

Putting a propeller on a cart with wheels, as shown in Figure 26-2, propels the cart forward (or backward if turning the other way).

What would you expect to happen if a sail is put in front of the propeller to catch the air, as shown in Figure 26-3? Some people would say the cart will move faster because the force from the fan will "push" the cart. However, what we find is this: with the sail in place, the cart does not move as it did without the sail. This is a surprising result for many people seeing this for the first time. The reason for this is, without the sail, the equal and opposite reaction of the propeller causes the cart to move forward. However, with the sail in place, the force of the propeller balances the reaction force. As a result, there is no net force and the cart does not move.

Figure 26-3 *With a "sail" the cart does not move.*

The Point

Linear momentum is conserved in the absence of external forces. For every action, there is an equal and opposite reaction.

Figure 26-2 *Without a "sail" the fan pushes the cart.*

Project 27
Newton's third law. Bottle rockets. Why do they need water? (Sir Isaac Newton in the passenger's seat.)

The Idea

In this experiment, you launch a 2-liter soda bottle into the air. Your fuel is water, which is propelled downward by air pressure forcing the rocket upward. This experiment is a good illustration of Newton's third law and the law of conservation of momentum, and it lends itself to a nice, friendly, competitive "space race."

What You Need

- 2-liter soda bottle (water bottles are not necessarily capable of sustaining internal pressure, as soda bottles are)
- nose cone fabricated from a cardboard party hat or a cone formed from poster board and tape
- cardboard for fins
- glue gun or tape
- water
- hard rubber stopper that just fits the top of the bottle (the stopper should be snug enough to seal the bottle while it is being pressurized, but not oversized to the extent that it prevents the bottle from launching)
- bicycle pump with a one-way valve (or an electric pump or compressor)
- optional: support to serve as a "launch pad" for the bottle (for instance, made from a tripod built from PVC pipe sections). See Figure 27-1.

Method

Build the rocket

1. Slide the open end of the bottle over the vertical rod of a ring stand for easier assembly.

2. Use the glue gun to attach fins to the rocket (remembering that the flat side of the bottle is the top of the rocket). Be careful not to apply excessive heat, which could melt a hole in the bottle.

3. Attach a nose cone to make the bottle more aerodynamic. Use poster board or a cone-shaped party hat.

Figure 27-1 *Bottle rocket ready for launch.*

4. Fill the bottle from about one-quarter to one-third full.

Assemble the launcher

You can do this in several ways. If you are planning many launches, you may want to go for something more elaborate. The basic parts are:

1. An air pump or compressor.

2. A one-way valve: The simplest way to do this is to insert a needle (available at any sporting-goods supply store) used to inflate footballs and basketballs through the stopper. With this method, no release mechanism is needed because the rocket will take off as soon as enough pressure builds up to overcome the force holding the stopper in the bottle.

3. A release mechanism: A metal "claw," which holds the bottle in place until the pressure builds to a certain level, allows a greater pressure to build up in the bottle. This can be mounted on a wooden or PVC tripod structure. You can also hold this in your hand, but be prepared to get wet as the "fuel" surges downward from the bottom of the rocket.

Launch the rocket

1. Insert the stopper into the bottle.

2. Secure the bottle onto the launcher. Move the holding mechanism into place (or hold it if that is what you are doing).

3. Pressurize the bottle. The maximum air pressure should not go above 80 to 100 psi (pounds per square inch) to avoid bursting the bottles.

4. Use a string to remotely release the release mechanism.

Expected Results

The rocket will ascend vertically. The upward leg path can take as long as about 4 seconds corresponding to a maximum height of more than 75 meters (over 250 feet).

Why It Works

The air pressure forces the water downward with a high velocity. The mass of the water times the velocity of the water represents the downward momentum of the water. Conservation of momentum requires an equal momentum upward that is applied to the mass of the bottle, which acquires a velocity to take it upward. Another way to say this is the action of the downward force of the water is counterbalanced by an equal and opposite reaction that drives the bottle upward.

Other Things to Try

Bottle rockets can be made more elaborate by adding fins. A parachute, made of the clear plastic used by dry cleaners, can be added to keep the rocket in the air for a longer time or to release a payload consisting of a tennis ball or other object.

You may want to see the *Mythbusters* episode, where they explore the use of bottle rockets to propel a person. Note, for safety reasons, they confined their efforts to dummies.

The Point

This is another example of conservation of linear momentum and Newton's third law.

Figure 27-2 *Sir Isaac Newton's laws of motion describe the motion of bottle rockets, satellites and planets.*

Project 28
Pushing water. Birds flying inside a truck.

The Idea

Newton's third law states that for every action (force), there is an equal but opposite reaction (force). This project illustrates how this concept can be applied to a particular physical situation. The outcome may be different than what many people expect.

What You Need

- 1 ping-pong ball attached to a string
- 2 beakers (or jars) filled with enough water to immerse the ping-pong ball
- balance scale
- counterweights

Method

1. Set each of the beakers on the opposing pans of the scale and establish a balance, as shown in Figure 28-1.

2. Predict what will happen when the ping-pong ball is lowered into the beaker of water. Will the side with the ping pong ball

   a. Rise?

   b. Fall?

   c. Remain balanced?

3. Lower the ping-pong ball into the beaker and observe what happens.

4. Remove the ping-pong ball. What is the effect on the balance?

Expected Results

Lowering the ping-pong ball into the beaker forces that side of the balance down, as shown in Figure 28-2.

When the ping-pong ball is withdrawn, the balance is restored.

Why It Works

There is a buoyant force on any object immersed in water (or partially immersed in water such as a floating ping-pong ball). For every action there is an equal and opposite reaction. In this case if the action (the buoyant force) is up, the reaction must be down, causing the observed effect.

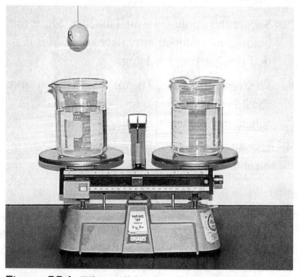

Figure 28-1 *What will be the effect of a floating object? Will it tip the balance?*

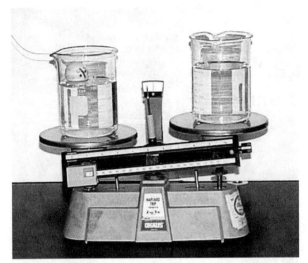

Figure 28-2 *Newton's third law: The bouyant force pushes up. The opposite reaction pushes the scale down.*

Other Things to Try

This is similar to the enigma: if birds are in a truck, will the truck weigh less if the birds are flying, instead of at rest on the floor of the truck bed? It turns out that the force exerted by the birds' wings exerts the same downward pressure on the truck bed as the weight of the birds at rest. (As with the previous experiment, this is also addressed by a *Mythbusters* episode.)

The Point

This experiment shows how a reaction force is established by Newton's third law.

Project 29
Slipping and sliding.

The Idea

This project compares the amount of friction developed by various common substances. It also shows a simple way to measure the amount of friction.

What You Need

- book
- coin
- ice cube
- rubber eraser
- protractor

Method

1. Line all three objects up in a straight line on the book.

2. Slowly lift the book.

3. As you raise the book, note the angle where each of the objects just begins to slide.

4. Break the rubber eraser into various pieces of different areas.

5. Line up the eraser pieces and determine the sequence in which they slide.

6. Break the ice cube into various sized pieces in different areas.

7. Line the eraser pieces up and determine the sequence in which they slide.

Expected Results

The ice goes first, then the coin, and then, finally, the eraser.

For a given material, the contact area between the sliding surfaces does not significantly affect the force of friction.

Why It Works

The surface of each material is characterized by a different amount of friction. It would be nice if this was called the surface's "slipperiness." However, it goes under the more prestigious name of *coefficient of friction*, which is the amount of frictional force a surface imposes on an object compared to its weight (on a horizontal surface). There is a frictional force to get something going (*static friction*) and a force to keep something going (*kinetic friction*). The frictional force depends on the coefficient of friction and the (horizontal) weight of the object. It does not (to first order) depend on the contact area.

The condition for the object to slide is that the tangent of the angle equal the coefficient of (static) friction. By finding the angle where the object just begins to slide, you can find the coefficient of static friction by simply taking the inverse tangent. (This is the key on your calculator that says *tan⁻¹*, *arctan* or *atan*.) See Figure 29-1.

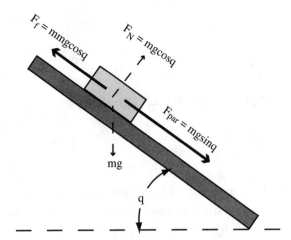

$F_f = \mu mg\cos q$

$F_N = mg\cos q$

$F_{par} = mg\sin q$

mg

q

Figure 29-1 *Forces on an incline.*

Other Things to Try

Friction applies a force that puts the brakes on motion. The amount of friction between two surfaces is characterized by something called the coefficient of friction, which is represented by the Greek letter μ (mu pronounced "myoo"). On a horizontal surface, the force exerted by friction is equal to the weight of the object, multiplied by the coefficient of friction. There are two types—static friction, which must be overcome to get something going, and kinetic friction, which must be overcome to keep something going.

In this project, the objects slide down the ramp if the tangent of the angle is greater than the coefficient of static friction for that object on the book. This condition can be turned around and, if the sliding angle is found, the coefficient of friction can be easily and simply determined.

A follow-up along a similar theme is to predict whether an object can slide along a surface without toppling. A book sliding front side down on a smooth table will not have stability problems. But a can of tomato juice sliding upright across a rough wooden floor may be another story. Try this with several cylinders with the same diameter, but with different heights. Sections of cardboard tubes or plastic pipe sections are good to test this. The condition for sliding without tipping is that the coefficient of (kinetic) friction be less than the ratio of the diameter of the cylinder to its length.

You could also design an experiment to study the effect of increasing the weight, surface area, and velocity of motion. You will find the weight of the object is the only significant variable and (surprisingly, for many people) the contact area is (almost entirely) insignificant.

The Point

Friction is a force that opposes motion. On a horizontal surface, the amount of friction depends on the weight of an object pressing it in contact with that surface, and the coefficient of friction.

Project 30
Springs. Pulling back. The further you go, the harder it gets.

The Idea

The force exerted by a spring, unlike any of the forces we have encountered so far, is not constant. It continuously varies. The further you pull the spring, the harder it pulls back. This relationship is known as *Hooke's law*. Because of this, springs have the capability to keep going back and forth until friction (eventually) slows them down.

What You Need

- various springs
- mass set (or a spring scale)
- metric ruler

Method

Measure the spring constant of a spring by following these steps:

1. Suspend the spring from the support.

2. Locate the distance from the bottom of the spring just hanging under its own weight. This is called the *equilibrium point*.

3. Hang a mass from the spring. The mass should be chosen so it increases the length of the spring by no greater than about 50 percent.

4. Measure the distance (in centimeters) the bottom of the spring is pulled below the equilibrium point. See Figure 30-1.

5. Find the force. If your reading is in grams, convert it to newtons by dividing the mass in grams by 1000 to get kilograms, and multiplying by 9.8 to get force in newtons. It may be easier to do this using a spring balance to get the force to pull the spring a certain distance. Many spring balances are calibrated directly in newtons, so in that case, there is no need to convert the force into newtons.

6. The spring constant can be determined by dividing the force by the distance according to the equation: $k = -F/x$. (Note: the negative sign accounts for the fact that the force and the extension are in opposite directions. If you pull up, the distance is positive, but the

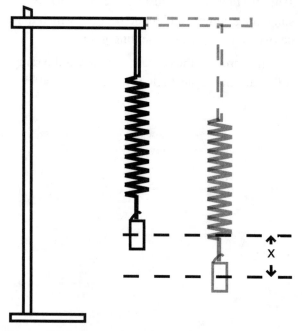

Figure 30-1 *Determining how stiff a spring is by measuring the spring constant.*

force is negative. Regardless of how you do it, the spring constant is *always positive*.)

Expected Results

The stiffer the spring, the higher the spring constant. As an example, if it takes 10 newtons of force to stretch a spring by 1 centimeter, the spring constant would be $k = 10N/1cm = 10N/cm$. It would take 20 newtons to stretch that same spring by 2 centimeters. The spring constant should prove to be constant and establish a linear relationship between force (F) and extension (x). Real springs have a linear range over which Hooke's law is a reasonable approximation. If you stretch too far, however, you will go out of the linear range and it usually takes an initial force to bring the spring into its linear range.

Why It Works

Over most of its range, the force exerted by a spring is directly proportional to the amount it is displaced from its equilibrium position. The spring constant is a way to characterize how much a particular spring exerts for a given displacement from equilibrium.

Other Things to Try

A more accurate value for the spring constant can be determined by taking several readings and plotting force versus displacement, and then finding the slope of the line.

The Point

The force exerted by a spring is proportional to the distance the spring is stretched. The further you pull, the greater the force the spring exerts in the opposite direction.

Project 31
Atwood's machine. A vertical tug of war.

The Idea

The Atwood machine illustrates some aspects of force and acceleration. Like an incline, the Atwood machine slows acceleration down to a measurable and observable amount. This project shows how the Atwood apparatus can be used to study acceleration.

What You Need

- pulley
- support for pulley, such as a ring stand
- string
- various masses

Method

1. Set up the apparatus with each of two masses attached to string and suspended over a pulley, as shown in Figure 31-1.

2. Release the masses and observe/measure their motion.

3. As in previous experiments, the acceleration of the masses can be measured using the stopwatch method (using $a = 2d/t^2$) or determining the acceleration using a motion sensor. (Different combinations of masses can also be compared and ranked visually without detailed measurements.)

Expected Results

The greater the difference between the two masses, the greater the acceleration.

The greater the combined masses, the smaller the acceleration.

The acceleration for the two masses is:

$$a = \frac{(m1 - m2)g}{(m1 + m2)}$$

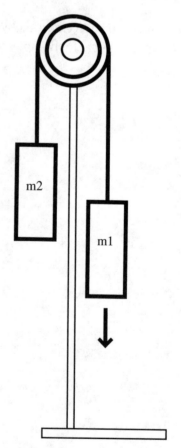

Figure 31-1 *Atwood's machine.*

If m1 is the larger mass, the acceleration is in the direction of the larger mass going down.

Why It Works

The force on the system is $F = g(m1 - m2)$. According to Newton's second law, this equals the total mass times the acceleration. Because the total mass is $(m1 + m2)$, we can derive the previous expression for acceleration.

Other Things to Try

Once you get the idea of this, try it at an incline, as shown in Figure 31-2.

Predict what angle or combination of masses will result in equilibrium.

Because friction helps establish stability, a window is around the predicted conditions that will also result in equilibrium. This can also be done with each of the two masses sliding on an incline.

The Point

An Atwood machine demonstrates the principles of Newton's second law. The net force on the masses causes the total mass of the system to accelerate.

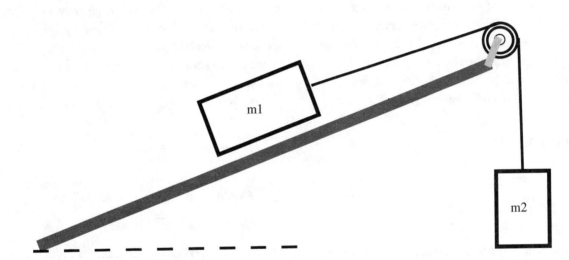

Figure 31-2 *Atwood's machine on an incline.*

Project 32
Terminal velocity. Falling slowly.

The Idea

Shortly after jumping from an airplane, skydivers reach a steady velocity instead of constantly accelerating as do other more streamlined objects subjected to Earth's gravity. When this happens, the skydiver falls at a terminal velocity that is nearly constant. This is fortunate because, once the parachute opens, it is much easier to slow the skydiver's fall. Had the skydiver been a rock in a vacuum without the benefit of air resistance, it would reach a much higher velocity.

What You Need

- coffee filter
- meterstick
- stopwatch
- coin, book, or another compact object
- optional: motion sensor

Method

1. Drop a coffee filter from a measured distance.
2. Compare the time it takes to fall with a coin or a book.
3. Compare the distance versus the time graph generated by a motion sensor for each of the two objects.

Expected Results

Not only will the coffee filter take longer to descend, but more significantly, it falls at a steady velocity. The coin, typical of other objects in free-fall, accelerates as it falls and will have an ever-increasing velocity. The following shows distance versus time and velocity versus time graphs for these two objects. Notice the falling book continues to accelerate as it falls. This is indicated by the curved shape of the position-time graph and the positive slope for the velocity-time graph.

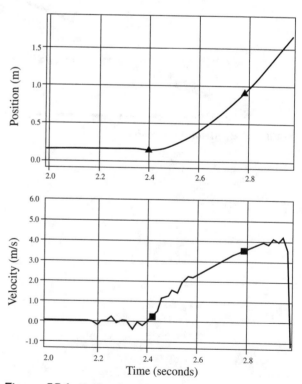

Figure 32-1 *Falling book.*

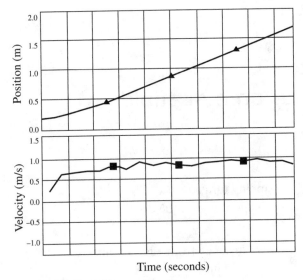

Figure 32-2 *Falling coffee filter.*

The coffee filter falls with constant velocity. The position-time graph is a straight line and the velocity-time graph is essentially a horizontal line, indicating a constant velocity during the descent.

Why It Works

When a falling object encounters significant air resistance, the faster the object falls, the greater the force opposing its descent. So, the more gravity tries to pull the object down, the more determined the air resistance is to oppose gravity. As a result, equilibrium is established with the object falling at a constant terminal velocity.

Other Things to Try

Compare the descent of bottle rockets (described in Project 27) with and without parachutes.

The Point

Free-fall is different than an object subjected to drag forces.

An object in free-fall accelerates with a constant rate of 9.8 m/s^2. An object subject to a drag force does not accelerate, but reaches a steady constant velocity, called the *terminal velocity*.

Project 33
Balancing act. Painter on a scaffold.

The Idea

A scaffold is built from a board placed across a base without anything holding it down. How far from the edge of the board can a painter stand without tipping the board? This experiment investigates the condition for stability called *static equilibrium*.

What You Need

- section of a 2" × 4" block about 6" long
- meterstick
- 20 g mass

Method

1. Set the block on the table. This can be either with the 2" edge parallel to the table or the 2" edge perpendicular to the table. Each case gives a different result.

2. Measure the mass of the meterstick.

3. Lay the meterstick over the block, as shown in Figure 33-1, with the 50-centimeter mark of the meterstick centered over the middle of the block.

4. Predict how far the 20-gram mass (the "painter") can be placed from the center without tipping the meterstick, as is the case shown in Figure 33-2.

5. The principle to use is that the torque trying to tip the "scaffold" must not be greater than the torque that holds it in place. Here are the formulas:

Tipping torque
$< $ supporting torque d_1w_1
$< d_2w_2$

d_1 = distance from edge of block to center of 20g mass

w_1 = weight of block

d_2 = distance from edge of block to the 50 cm mark of the meterstick

w_2 = weight of the meterstick

A 2" × 4" block has actual measurements of 1½" × 3½" (or 3.8 cm × 8.9 cm). (The 3.8 cm side is the height and the 8.9 cm side is the width of the block.) The distance, d_2, is one-half the supporting edge. This would be 4.45 cm (with the width of the block along the table) or 1.6 cm (with the height of the block on the table).

6. Try it with other masses.

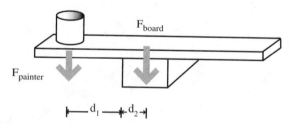

Figure 33-1 *How far can the "painter" move toward the edge of the board?*

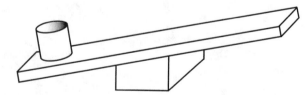

Figure 33-2 *Here the "painter" has gone too far.*

Expected Results

For a 90-gram mass meterstick balanced on top of a nominally 2" × 4" block, the following table shows the maximum distance the painter, m2, can go without toppling the meterstick.

	Mass of "painter"	Maximum distance from edge before "scaffold" topples, d2
Block width along	10 g	40.1 cm
table (d1 = 4.45 cm)	20 g	20.0 cm
	50 g	8.0 cm
	100 g	4.0 cm
Block height along	10 g	14.4 cm
table (d1 = 1.6 cm)	20 g	7.2 cm
	50 g	2.9 cm
	100 g	1.4 cm

Why It Works

The amount of mass carried at a point of support is the result of a torque generated around the pivot point. In this case, the spring scales form a pivot point. The greater the mass supported, and the further from the pivot point, the greater the torque.

Other Things to Try

How weight is distributed

Place two bathroom scales on the floor separated by the length of the board. Set a stiff board about 8 feet long over each scale. Adjust the scales to read zero, to eliminate the effect of the weight of the board. Predict and measure the reading directly over the scales, in the middle, and at arbitrary positions in between.

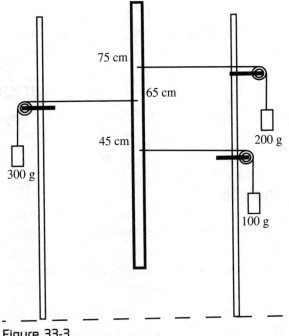

Figure 33-3

Vertical static equilibrium

Assemble the apparatus shown in Figure 33-3. Based on balancing clockwise and counter-clockwise torque, develop other combinations that establish equilibrium. This is based on a demonstration found on the U.C. Berkley Physics Lecture Demonstration website http://www.mip.berkeley.edu/physics/noteindex.html (item: A+60+0).

The Point

Static equilibrium reflects a balance of forces that results in a collection of objects remaining stable and stationary. The condition for static equilibrium is that the sum of the force and the sum of the torques on an object is zero.

Project 34
Hanging sign.

The Idea

You are hanging a sign for your café that weighs 50 pounds. You have one cable that is 4 feet long and another cable that is 5 feet long. Which one supports more of the weight?

Angles complicate how forces are distributed. This project explores a simple situation similar to hanging a sign with two different length cables.

What You Need

- 2 spring scales
- mass—(should give close to a full-scale reading in the vertical position on your spring scales)
- string
- 2 ring stands with clamps or comparable support
- key ring
- protractor

Method

Symmetrical sign

1. Place the ring stands about 18 inches apart.

2. Cut two equal-length sections of string. The strings should be 8 inches, leaving a couple of inches on each side for attaching to the support and the weight.

3. Set the spring scales to read zero (in the position they are being used), with no weight hanging from them.

4. Hook both of the spring scales to each of the ring stands.

5. Attach each string—one side to the key ring and the other side to the clamp on the ring strand.

6. The apparatus should be as shown in Figure 34-1.

7. Hang the mass on the key ring.

8. Record the reading on each of the spring scales.

9. Repeat, using different masses and different string lengths.

Asymmetric sign

1. Repeat the previous steps, using different string lengths.

2. Based on your evaluation, can you answer the question posed at the beginning of this section?

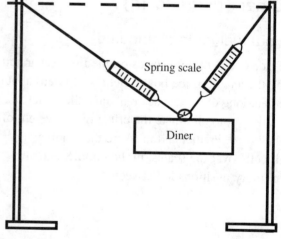

Figure 34-1

Expected Results

In the case of the symmetric supports, the two scales will read the same.

If the strings are different lengths, the *shorter* of the two strings will bear *more* of the weight.

Why It Works

The force in a cable is the combination of the various forces present. The overall force depends on how large each of the forces is and its direction. The method of combining these forces is called *vector addition*.

Other Things to Try

Tension is the force in a rope or cable that can change direction without losses by going around a pulley. How do you think the force in each of the cases in Figure 34-2 compares? Test it out with weights and pulleys.

Each spring scale should read 9.8 newtons, which is the amount of force exerted by gravity on a 1 kg mass.

Tug of war. It is almost impossible to pull a rope supporting a moderate weight tight enough to be perfectly horizontal. The experimental setup is shown in Figure 34-3. A gallon milk container filled with water makes a good 4 kg mass to try this with.

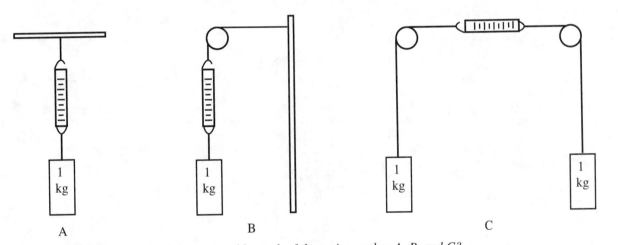

Figure 34-2 *How much force is measured by each of the spring scales: A, B, and C?*

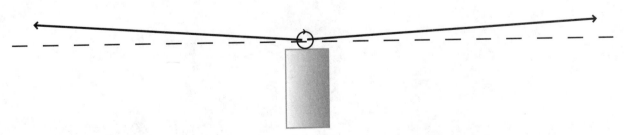

Figure 34-3 *With a 4 kg mass, it is almost impossible to pull the rope perfectly horizontal.*

To bring the rope to within 5 degrees of horizontal with a 4 kg weight in the middle, you need to pull with a force of 560 newtons (or over 125 pounds). To bring the rope to within 1 degree of horizontal, you need to apply a force of about 2300 newtons (or over 500 pounds).

The Point

This exercise shows how a force can be broken down into or resolved into a pair of component forces. The downward force from the suspended sign is supported by cables of various lengths in various directions. This is the type of thinking civil engineers apply on a daily basis—to oversimplify, how strong something (like a bridge cable) needs to be to support a load (like the road surface with cars).

Project 35
Pressure. Imploding cans.

The Idea

How powerful is air pressure? We can answer this by observing what happens when we remove air from where it is normally found.

What You Need

- empty soda cans
- or a larger rectangular can with a screw-on cap, if you can get one. (These are used for cooking oil or paint thinner, and they can be purchased to do this experiment.)
- hotplate
- beaker
- water
- oven mitt or beaker tongs
- pail of cold water

Method

1. Rinse the can, so it is reasonably clean.

2. Put a small quantity of water (2 tablespoons) in the can.

3. Place the can on the hotplate and keep it there until the water boils and steam comes out of the can (Figure 35-1).

4. Carefully remove the can from the hotplate using the mitt or tongs.

5. Quickly immerse the soda can, with steam still evolving, top side down in the water. Observe what happens (Figure 35-2).

Figure 35-1 *A small amount of water in the can is heated.*

Figure 35-2 *Lower air pressure in the can causes it to be crushed.*

95

6. Remove the rectangular can from the hotplate. With steam still coming out of the can, screw on the cap. Wait until it cools in the air or facilitate the cooling with water or ice. (If you are doing this as a demonstration, this may take a while and a sense of drama can be created by pretending that nothing is happening and going on to the next experiment.)

Expected Results

The soda can will be crushed almost instantaneously. The rectangular can may take a few minutes; it gets crushed slowly (as if by a protégé of Darth Vader using the Force).

Why It Works

Air exerts a pressure of 14.7 pounds on every square inch. As the steam in the cans condenses, the air pressure inside the can drops and there is a difference in pressure between the inside and outside of the can. This is primarily the result of the change in state from vapor to liquid and to a lesser degree from the contraction of the gas in the can as it cools. This means a soda can (6.5 inches high and 3 inches in diameter) has a force of over 900 pounds pressing down on its sides!

Other Things to Try

Try this with a 55-gallon drum as shown in Figure 53-3. Use a vacuum pump connected to the drum through a valve to create the pressure difference. This may take some time, but it will be worth the wait. You may want to warn the people you work with that the boom they are about to hear does not require the emergency response team to be sent in.

The Point

Air pressure is substantial, exerting a force of nearly 15 pounds for every square inch that it is in contact with.

Figure 35-3 *Air pressure did this.*

Project 36
Pressure. Supporting water in a cup.

The Idea

Which weighs more: an empty cup or a full cup? Obviously, the full cup. So, if you turn a cup upside down with a flat cover, which has the better chance of being supported: the (lighter) empty cup or the (heavier) full cup? Many people find the outcome surprising.

What You Need

- cup or beaker with a flat, circular, smooth top. (If the beaker has a spout, it should be flat enough so it can make contact with a card at all points along its top surface.)
- water to fill the cup
- flat, stiff, lightweight square, large enough to cover the entire top surface of the cup. The square should not become waterlogged. Cardboard is not the best choice. Foamboard stands up to water better. Index cards can work, but you must be careful not to let them flex and break the seal with the cup.

Method

1. Start with the empty cup. Cover the empty cup with the square. Turn it upside down and observe what happens (Figure 36-1).
2. With the cup still empty, moisten the top surface of the cup to make it more sticky. Cover the cup and turn it upside down. What happens?

3. Now fill the cup with water to the brim. It won't hurt to have it go above the surface of the cup or to allow some water to spill out. Cover the cup with the square. Invert and observe what happens.

Expected Results

The square will fall off with the empty cup, even with the benefit of the surface being sticky. The water in the full cup, however, will be held in place by the square (Figure 36-2).

Why It Works

The water in a beaker 5 centimeters in diameter and 12.7 inches tall has a mass of 250 g (0.25 kg) and weighs 0.55 pounds (or 2.5 newtons). The air

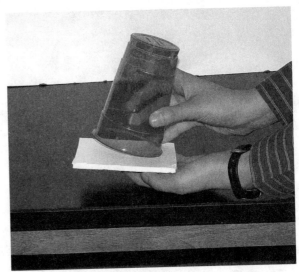

Figure 36-1 *With the cup empty, there is nothing to hold the square onto the cup.*

Figure 36-2 *The force exterted by air pressure is greater than the weight of the water.*

pressure on the 5 centimeter (roughly 2 inch) diameter circle is over 3 pounds. The air pressure is far greater than the weight of the water in the beaker. The empty beaker has the same pressure inside and outside the cup, so the air pressure is balanced and the weight of the cup causes it to fall. The adhesion of the square to the cup is clearly not enough to make up the difference.

Other Things to Try

Calculate how tall the glass can be with the card supported by air pressure. (The density of water is 1 g for every cubic centimeter and 1000 g is the equivalent of 2.2 pounds.) Atmospheric pressure will support a column of water 34 feet high. Since mercury is more dense than water, atmospheric pressure will support a column of mercury about 30 *inches* high. The exact height varies with local air pressure and provides a way to measure changes in air pressure.

The Point

Air pressure is large compared to the pressure exerted by the weight of a cup of water.

Project 37
Pressure. Sometimes the news can be pretty heavy.

The Idea

How much pressure does the atmosphere exert on a sheet of newspaper? As with the previous experiments, the force exerted by air pressure can be surprisingly powerful.

What You Need

- section of newspaper
- table
- piece of wood about 12 to 24 inches long and roughly 1 to 2 inches wide. (The wood should be thin enough so it can be readily snapped in half by someone who is not a black belt. It should also be stiff enough so it will break rather than flex if stuck. A ruler that you are willing to dedicate to the cause of science usually works.)

Method

1. Place the piece of wood on the table, extending approximately one-half its length.

2. Place a few layers of the newspaper over the wood. Lay it out so it is as flat as possible. Remove any "air pockets" that you can under the paper and make sure the edges are flat.

3. Using your best Maxwell Smart karate chop, strike the wood. Hit it hard enough to break the wood. Show it no mercy (Figure 37-1).

Expected Results

Many people would expect the wood to push the paper up and throw the paper partway across the room. However, if the paper is properly sealed over the wood, striking the wood results in the wood being pinned to the table and breaking apart as if it were clamped to the table.

Why It Works

Let's say you have a 1-inch wide ruler that extends 10 inches under the paper. This means that *147 pounds of air pressure* is pressing down on the ruler. About the same as a medium-sized person standing on the paper holding down the ruler. Air pressure is *that* strong. If this doesn't work, it isn't because of insufficient air pressure. The ruler may not break if: air leaks under the paper from the edges, the wood is too flexible to break, or the wood is too thick to break. It is unnecessary, of course, to actually break the ruler to demonstrate the strength of air pressure on the paper.

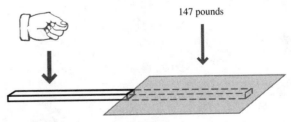

147 pounds

Figure 37-1 *Breaking a board with only air pressure holding the other side down.*

Other Things to Try

Air pressure on paper can also be observed pressing down on the pages of a book. After interleaving the pages of two similar books, it will be very difficult, if not impossible, to pull the two books apart. This is not the result of the friction of the pages, but it is a direct effect of the air pressure holding the pages together.

The power of air pressure can also be demonstrated using a suction cup such as those used to pop out minor dents in car side panels. To do this you will need an object with a smooth surface. One good example is a laboratory stool.

1. Attach the suction cup to the top surface of the stool, as shown in Figure 37-2. Make sure the suction cup seals to the top surface of the stool.

2. Pull up on the suction cup.

The suction cup should be able to lift an average-sized stool up off the ground. There is a common misconception that a vacuum somehow pulls or "sucks" objects to it. This is not the case. Suction cups work because of a *difference* in air pressure between the outside of the suction cup and the little air trapped under the suction cup. The pressure on a suction cup that is 4 inches in diameter (assuming a perfect seal) would be greater than 150 pounds.

Figure 37-2

The Point

Air pressure exerts a force on a surface in proportion to its area.

Project 38
Archimedes's principle. What floats your boat?

The Idea

Does iron float? Clearly a cube of iron, which is much denser than water, will sink. Then, how is it possible for a boat made of iron (or iron alloy) to float? In this experiment, you investigate the forces that counteract the force of gravity to allow objects that are denser than water to float.

What You Need

- "boat": a shallow plastic cup (such as a ¼ pound coleslaw container). You can also use a piece of wood as your boat.

- "lake": a plastic tray or fish tank filled with water deep enough and wide enough to float the "boat"

- "cargo": small weights, pennies (each penny has a mass of 2.7 g)

- 100 mL graduated cylinder

Method

1. Measure the volume of the boat. You can do this by geometry, if you are so inclined, or you can do it by filling the cup with water and measuring the amount of water to do this.

2. The number of grams of water that occupy the volume of the boat equals the number of grams of cargo it can carry just before it sinks.

3. Test this by adding the amount of weight you predicted. Don't forget to include the weight of the boat itself. If you use pennies, count each as 2.7 grams (or weigh them). As you add weight, be careful not to tip the boat or you will capsize it prematurely.

4. Compare your prediction with the amount of cargo your boat could actually carry. See Figure 38-1.

5. We are taking a slight liberty here for the sake of clarity by focusing on the mass. What holds the boat up is a buoyant *force*, which is measured in newtons. The buoyant force equals the *weight* of the water (also measured in newtons) displaced by the floating object.

Expected Results

As a rule of thumb, for every 1 mL that an object is held submerged below the surface of the water, there is a buoyant force capable of supporting 1 gram of mass.

An equivalent way of expressing this is an object will float if the density of the entire boat,

Figure 38-1 *The buoyant force equals the weight of the boat plus its cargo. This can be predicted by determining the weight of the displaced water.*

considering its entire volume, is less than the density of the same volume of water. The density of water is 1 gram/cubic centimeter or 1000 kilogram/cubic meter.

Why It Works

The buoyant force on an object is given by the weight of the fluid it displaces.

The buoyant force exerted on a floating object equals the volume of that object that is submerged (m^3) times the density of water (1000 kg/m^3) times the gravitational acceleration (9.8 m/s^2).

Other Things to Try

1. Take a weight of known volume. Either calculate the volume using geometry or determine how much water it displaces. Measure the weight on a spring scale in the air. Immerse the weight in the water. How is the weight affected by being immersed in water? How does this compare with the weight of the water that would fill the volume of the submerged object?

2. Predict how much weight a boat can support based on the weight of the water that it can hold. Test your prediction.

3. Predict where a float line will be based on the volume contained below the float line set equal to the weight to be added to the boat.

The Point

A buoyant force is exerted on an object that is floating or submerged in a liquid. An object will float if it is less dense than the liquid it is floating in. The buoyant force exerted on an object floating in water equals the *weight of water* that would occupy the *volume* of the object that is *underwater*.

Project 39
Cartesian diver.

The Idea

An object is barely floating in a bottle. In this project, you control whether it floats or sinks, just by applying pressure on the side of the bottle.

What You Need

- 2-liter plastic soda bottle with a reclosable cap
- water
- 1 medicine dropper
- a small object such as a paper clip to fine-tune the weight of the medicine dropper

Method

1. Partially fill the medicine dropper with just enough water to establish *neutral buoyancy*, which is a condition where the medicine dropper does not sink, but also does not float in water. Test and adjust until the right amount of water is in the medicine dropper. Use the paper clip to increase the weight of the medicine dropper.
2. Put the medicine dropper in the soda bottle.
3. Fill the bottle to the top with water.
4. Cap the bottle securely.
5. Observe what happens as you apply pressure to the sides of the bottle. What happens as you relieve the pressure?

Expected Results

When the side of the bottle is squeezed, the diver descends.

When the pressure is released, the diver returns to the surface.

Why It Works

Pressure on the side of the bottle is transmitted to the medicine dropper. The pressure reduces the volume of the medicine dropper. The buoyant force depends on the size of the medicine dropper. A smaller-volume medicine dropper has a smaller buoyant force and sinks.

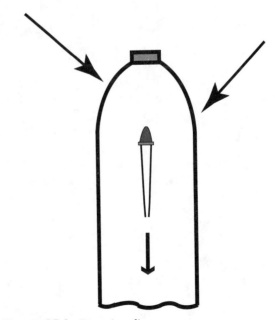

Figure 39-1 *Cartesian diver.*

103

With the pressure reduced, the volume of the medicine dropper increases, leading to greater buoyant force. Greater buoyant force causes the medicine dropper to rise.

Neutral buoyancy occurs when the force of gravity equals the buoyant force resulting in equilibrium in the vertical direction. This is shown for a SCUBA diver in Figure 39-2. A diver would add weight to a belt as you did above with the paper clip to fine-tune the balance of forces.

Other Things to Try

Once you get the hang of this, you can put a "treasure" in the form of a small weight with a hook at the bottom of your bottle. You can then have your diver go down and try to recover the treasure.

The Point

Buoyancy depends on the volume of a submersed object. Pressure exerted externally reduces the volume of a submersed object, which, in turn, reduces the buoyant force.

Figure 39-2 *Vertical forces in equilibrium. Photo by Dr. Michael Dershowitz.*

Project 40
An air-pressure fountain.

The Idea

This is a nice attention-getting demonstration that produces a fountain-like spray in an inverted flask. If you are doing this as a demonstration—especially for younger children—be prepared to be asked to do it again.

What You Need

- ring stand with a small 2-inch diameter ring
- flask (250 ml works)
- 1-hole rubber stopper to fit the flask
- approximately 12-inch section of glass tubing that can be inserted into the stopper
- hotplate
- oven mitt
- beaker of equal or larger volume as the flask
- water (with food coloring optional)
- safety glasses

Method

1. Put on safety glasses.
2. Carefully slide the glass tube through the stopper, so approximately 1 inch protrudes through the narrow end of the stopper. Use proper techniques for handling the glass tubing (including wearing eye protection, protecting your hands as you push it through using a towel, and lubricating the edge with a bit of Vaseline, so you don't have to force it through the hole).

3. Insert the stopper into the flask.
4. Attach the ring on the ring stand high enough so the entire length of the tubing is supported about ½ inch above the base of the stand.
5. Fill the beaker close to the top with water. (Food coloring can temporarily stain your fingers.)
6. Assemble the apparatus, as shown in Figure 40-1. Make sure everything fits and is secure.

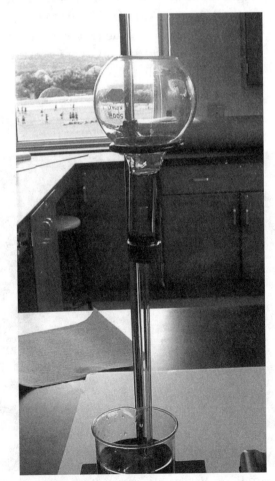

Figure 40-1 *Air pressure fountain.*

7. Take the flask out of the ring stand and remove the stopper with the tubing.

8. Put (a few tablespoons of) water into the flask. Set the stopper on a table.

9. Place the flask on the hotplate (Figure 40-2).

10. When the water starts to boil and the flask fills with steam (using the oven mitt), remove the flask from the hotplate and attach the stopper.

11. Quickly, but carefully (still using the oven mitt), reassemble the apparatus. One convenient way to do this is to note the position of the ring, remove the ring, and then place the ring over the collar of the flask. Then, with the stopper inserted, invert the flask, and (with the flask supported by the ring as it is transferred) reattach the ring on the stand. It wouldn't hurt to choreograph this a little bit before doing it (and have a second person help you). The idea is to do the transfer quickly (so you don't lose all of your steam), but *safely* (because you are working with glass and hot liquids). Placing an ice cube on top of the flask may further accelerate the process.

12. With the flask in the ring stand and the glass tubing close to the bottom of the beaker, observe what happens.

Expected Results

At first, the water starts to rise up the tube. This begins slowly at first. As the water works its way up the tube, it begins to pour into the flask. Once the water touches the interior of the flask, it begins to spray, forming a fountain that increases in intensity until the water is completely drawn out of the flask (Figure 40-3).

If positioned just right, the fountain ends in a gurgling effect. While many observers may

Figure 40-2 *Heat a small amount of water in the flask.*

Figure 40-3 *Air pressure causes the liquid to spray into the flask.*

expect the rise of the liquid up the tube, the surge of the fountain catches many people off guard.

Why It Works

As the steam inside the flask begins to cool, the air pressure inside the flask drops. This is primarily the result of the phase change of the steam from vapor to liquid water, which occupies a much smaller volume. The cooling air inside the flask also contracts, adding to the reduced pressure. Atmospheric pressure pushes down on the liquid in the flask, driving up into the glass tube. The cooler the flask gets, the lower the pressure. This process feeds on itself in an accelerating manner, producing the fountain effect.

Other Things to Try

The mechanism that drives the liquid up into the flask is the basis for what is known as a Torricelli barometer. Air pressure is measured by how high a column of water can be supported by air pressure with a vacuum in the flask. Mercury is used instead of water because standard air pressure can support a mercury column roughly 30 inches high, compared with a much-higher column for water. Because of potential difficulties in working with mercury in academic settings, it is probably best just to read about this one.

The Point

This project works because of the volume differences between vapor and liquid, and the force exerted by air pressure.

Project 41
Blowing up a marshmallow. Less is s'more.
Why astronauts do not use shaving cream in space.

The Idea

The next time you are sitting around the campfire cooking up a batch of s'mores, be sure to point out to your friends that a marshmallow is simply a colloidal suspension of air in a solid. Because the air in the marshmallow is in equilibrium with the atmosphere, the volume of the marshmallow is stable at standard air pressure. However, it's a different story if we disturb the equilibrium conditions by taking away the atmospheric pressure.

What You Need

- marshmallow
- bell jar
- vacuum pump

Method

1. Place a marshmallow on the base of the bell jar.
2. Assemble the bell jar and apply a vacuum.
3. Observe what happens.

Expected Results

The marshmallow grows in volume, as you can see in Figure 41-1.

Why It Works

The pressure of the air trapped in each marshmallow causes the marshmallow to expand

Figure 41-1 *Air pressure trapped in a marshmallow causes expansion of the marshmallow. (The vacuum jar pictured here was adapted from an apparatus used to show the dropoff in sound transmission as air pressure is reduced.)*

when the pressure outside the marshmallow is reduced.

Other Things to Try

Try this with shaving cream. It will increase in volume.

Try this with hot water, just under the boiling point. The water should begin to boil again.

The Point

Common objects depend on air pressure for them to maintain their physical shape and appearance.

Project 42
Relaxing on a bed of nails.

The Idea

Would you sit down on a bunch of nails that are sticking out of a wooden board with the sharp pointy sides upward? This is your chance to try. This is not nearly as painful as it may seem because the large number of nails spreads the force over a larger area.

What You Need

- 144 nails, 1½ to 2 inches long. Whatever length you use, make sure the nails are nearly all the same length
- piece of plywood 14 inches × 14 inches × ¾ inches thick (or larger)
- electric drill
- drill bit whose diameter is equal to or just slightly smaller than the diameter of the nails
- inflated balloon

Method

Assembling the bed

1. Draw evenly spaced lines at 1-inch intervals running parallel with each of the edges of the board.

2. Drill a hole at the intersection of each of the holes.

3. Insert the nails in the holes. They should be snug enough not to fall out. Some may have to be driven in with a hammer (Figure 42-1).

Testing the bed

1. Press the balloon on one of the nails on the corner.

2. Press (another) balloon in the center of the bed (Figure 42-2).

3. Place the nails on a chair and sit down on it.

Figure 42-1 *Bed of nails.*

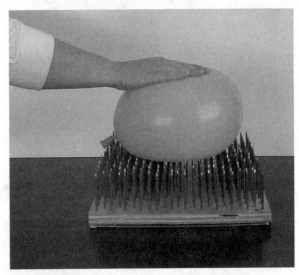

Figure 42-2 *The force is spread out over a large number of nails.*

Expected Results

With one nail isolated, the balloon will burst. However, with several nails in contact with the balloon, the balloon does not burst, even with substantial pressure applied. Sitting on the bed of nails is surprisingly painless.

Why It Works

Pressure is force divided by area. For a given force pressing down on the balloon, the pressure is much greater with the single nail than with the large group of nails. When you sit on the group of nails, the force is the result of your weight, but the pressure is spread out over the large number of nails.

Razor blades that use multiple blades apply the principle of spreading out the force exerted by any one blade over a large surface area.

Other Things to Try

Instead of just a seat, why not build a nice comfortable bed to sleep on?

You can also show how pressure can be spread out over a larger area by using several cups to support a person. Set up a board on a row of

Figure 42-3 *The person's weight is distributed over several cups.*

paper cups, as shown in Figure 42-3. Then, have a person stand on the board. If you have enough cups, you should be able to stand on the board supported by the paper cups. Some cups are stronger than others, so you may have to experiment to determine how many you need. Placing them every 2 inches or so, however, is a good place to start.

The Point

Pressure is force divided by area. The larger the area a force is applied over, the smaller the pressure experienced.

Project 43
Blowing hanging cans apart. What Bernoulli had to say about this.

The Idea

Here is a simple challenge: hang two empty soda cans from a string, separated by a few inches, and then blow them apart. Whether you use your lungs or a hair dryer, the result will be the same.

What You Need

- 2 empty soda cans
- 2 strings
- blow dryer

Method

1. Attach the strings to the can and hang from a support (such as a ring stand).

2. With the blow dryer (or your own breath if you are good at blowing out birthday candles), direct a stream of air between the cans. Avoid blowing so hard that the cans start bouncing around, which will only serve to confuse the issue.

3. Observe what happens (Figure 43-1).

Expected Results

Instead of blowing the cans apart, the air stream will drive them together. In fact, the harder you blow, the greater the force that moves the cans together.

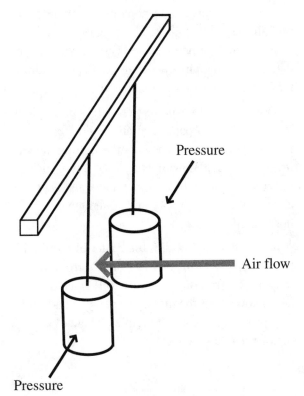

Figure 43-1 *Greater air velocity results in lower pressure to draw can together.*

Why It Works

What is going on here was explained by Bernoulli. A key consequence of the principle that bears his name is the faster air moves, the lower the air pressure. This is the mechanism for airplane lift. The contour of the airplane wing directs air above the wing at a higher velocity, resulting in a lower pressure above the wing. A spoiler on a race car does the same thing, but upside down. A sailboat takes advantage of

Bernoulli's principle by directing the air in front of the sail at a higher velocity. This produces a greater force on the wing and enables the sailboat to move at a greater velocity than the wind driving it.

Other Things to Try

A stream of moving air from a blow dryer keeps an object, such as a ping-pong ball, suspended in mid air. See Figure 43-2. This is another example of Bernoulli's principle. The faster moving air in the center results in a pressure gradient that draws the ball into the air steam above the blow dryer.

Another way to explore Bernoulli's principle is to take a sheet of paper and hold it horizontally in front of your mouth. The paper will droop down in front of you. Blowing across the top of the paper will reduce the pressure above the paper, causing it to defy gravity and straighten out horizontally.

If you place a dowel in the center of a roll of toilet paper and direct a blow dryer across the top, you can run out the entire roll! If anyone complains about the mess, just say you had to do it to prove Bernoulli's principle. (Then, of course, please clean up the mess.)

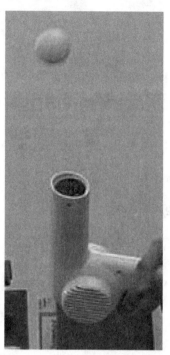

Figure 43-2 *Ping pong ball levitated by Bernoulli's principle.*

The Point

According to Bernoulli's principle, *moving air* results in *lower pressure*.

Project 44
Center of mass. How to balance a broom.

The Idea

Experience tells us that objects are more stable if their center of mass is lower to the ground. Based on that, you might think it would be easier to balance a broom with brush side down. This experiment lets you answer whether that is the case.

What You Need

- meterstick
- 2 books (physics textbooks are preferred, but English textbooks work almost as well)
- duct tape
- alternative: you can do this with an actual broom or any other object that has much of its mass concentrated at only one end. This

can be done with modeling clay attached to the end of a broom or pencil.

Method

1. Insert the meterstick between the two books, so an inch or two of the meterstick protrudes beyond the bottom of the book.
2. Secure the book to the meterstick.
3. Predict which end of the meterstick you should support to most easily balance it: the heavy end or the light end?
4. Support the meterstick on the heavy end.
5. Try this with the heavy end up and the light end supported by your hand.

Expected Results

One might say that with the mass at the bottom, the meterstick will be more stable. The logic is, like with a drag racer, placing the center of gravity at the lowest point possible results in the greatest stability. The results of trying this, however, reveal the opposite to be the case. It is easier to balance the meterstick with the weight at the top, not the bottom.

Why It Works

The reason for this unexpected behavior is that a small movement at the support end creates a greater torque with most of the weight located at the opposite end. This gives the person trying to balance the meterstick greater control. This

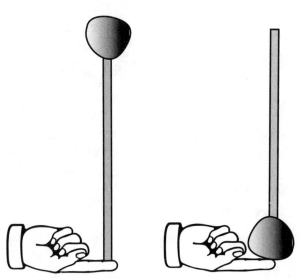

Figure 44-1 *Which is easier to balance?*

principle is used by tightrope walkers who carry a pole with a weight at the end to help establish balance.

Other Things to Try

Skyscrapers are often subjected to vibration when they're exposed to wind. Sometimes, adding mass to the structure can damp down the extent of the swaying. Apply the results of this investigation to determine whether it is more advantageous to add mass to the top floor or to the first floor of a skyscraper. Look up a specific example of how mass was added to the top, rather than the bottom, of the Sears Tower in Chicago.

The Point

Adding mass away from the pivot point increases the torque produced at the other end. This provides a greater degree of control to the end without the weight.

Project 45
A simple challenge. Move your fingers to the center of a meterstick.

The Idea

OK. Here is another simple challenge: Get a meterstick. Place one finger near the 15 cm mark and the other finger near the 65 cm mark. Move both fingers together at approximately the same velocity, so they meet *together* at the 40 cm mark. Is that asking too much?

What You Need

* meterstick

Method

1. Place the meterstick horizontally and hold with an outstretched finger from each hand.

2. Place your fingers near the 15 and 65 meter markings of a meterstick (Figure 45-1).

3. Move both fingers at roughly the same velocity, so they meet at the 40 meter mark.

Expected Results

This does sound simple enough but this is just about impossible for most people to do. You will find you can *only* move the finger furthest from the center (the one starting at the 15 centimeter mark) until both fingers are the same distance from the center. Then, they meet close to the middle (the 50 cm mark). See Figure 45-2.

Why It Works

A lot of physics is actually in this little investigation. The force is greater on the finger furthest from the center (because there is greater *torque* trying to rotate the meterstick in that direction). The greater the force, the greater the

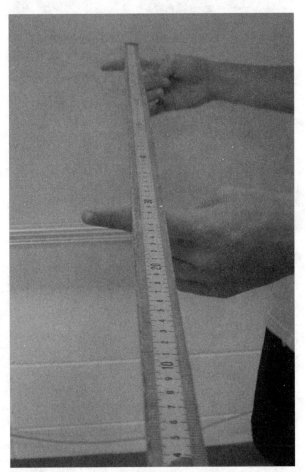

Figure 45-1 *The finger that has started on the 15 cm mark starts to move while the finger on the 65 cm mark hasn't moved at all.*

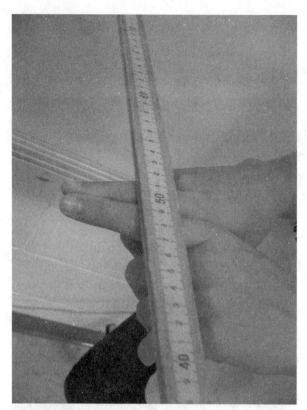

Figure 45-2 *Both fingers eventually meet at the 50 cm line. Not the 40 cm line.*

force of friction. This results in one finger being able to move much more easily than the other.

Other Things to Try

Try this with different starting finger positions. You may also need to convince any skeptics that one side of the meterstick does not have more friction than the other.

The Point

The further a weight is from a pivot point, the greater the force it exerts. Greater force between the surfaces in contact results in greater friction.

Project 46
Center of gravity. How far can a stack of books extend beyond the edge of a table?

The Idea

You have four equal books. Each is 10 inches long. You can stack them up any way you like. How far from the edge of the table can you place the far edge of the top book, so all four books still balance over the edge of the table? This can be done by intuition or analytically. It also makes a good competition activity.

What You Need

- stack of objects: bricks, blocks, books, or empty CD cases
- ruler

Method

1. Before you start, state your prediction. How far beyond the edge of the table will the four books go, so they balance without falling? See Figure 46-1.

2. Take the four books and arrange them, so the fourth object extends as far from the edge of the table as possible.

3. Repeat with any other number of books. This makes a good friendly competition to see who can produce the greatest overhang.

Expected Results

If you have four similar objects, 10 inches long, the maximum overhang will be just under 9.4 inches. In general, if you have four objects whose

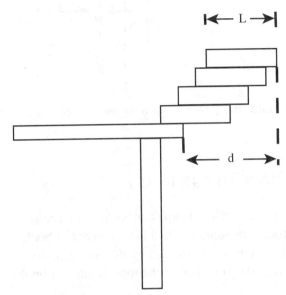

Figure 46-1 *How far can the books extend beyond the edge of the table?*

length is *L*, the maximum overhang is (just under) 0.94 × L.

Why It Works

The books (or other objects) will balance on the table if the center of mass for all the books lies over the table. If the center of mass is positioned over the edge of the table, the entire stack of books will topple.

Let's start with one book. The book balances with the overhang no greater than halfway. With a second book added, the equilibrium is maintained with the added book extending one-quarter of its length beyond the first. As books are added, the added distance that the entire stack can be pushed out is one-half the length of that of the previous book, as indicated in Figure 46-2.

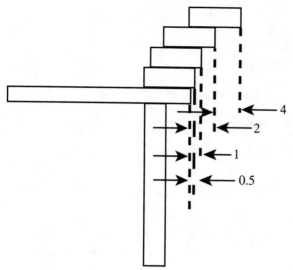

Figure 46-2 *Spacing for a maximum overhang.*

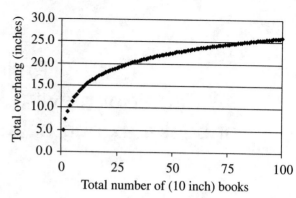

Figure 46-3

The total overhang is the sum of all the individual extensions. As an example, for 100 books, each of a length 10 inches, the maximum overhang would be 25.9 inches as pictured in Figure 46-3.

Other Things to Try

Now that you got past four books, how about 100? Extending this to a large number of books, the overhang is extended by the book length divided by one-half of the total number of books.

The Point

Equilibrium is maintained when the center of mass is centered over the area of support.

Project 47
Center of mass. The leaning tower of pizza.

The Idea

How far can an object tilt before it falls? Like the Leaning Tower of Pisa, the stability of a rectangular or cylindrical object depends on its shape. This experiment establishes a simple condition for stability of an object and explores the idea of center of mass.

What You Need

- cereal box
- pizza box
- 2 pencils
- tape
- string
- 2 nuts, large washers, or other matched attachable weights
- wooden board to use as an incline (roughly 3 ft × 4 inches × ½ inch, or 1m × 0.1m × 0.01m)

Method

1. Find the center of each of the rectangular faces of the box.

2. Start with the largest face first. Push one pencil through both sides of the box. The pencil should be roughly perpendicular to the surface it is pushed through.

3. Tie the string—one end to the pencil and the other end to the hangable weight.

4. Attach the other weight to the other side of the pencil as a counterbalance. You can use string if that makes this easier.

5. Tape the other pencil across the incline, somewhere roughly near the midpoint.

6. Place the box on the incline, so the downhill side of the box is in contact with the pencil taped to the incline. This pencil serves as a pivot point to force the box to rotate, rather than slide down, the incline.

7. Make your predictions. How far can you lift the incline before the box topples?

8. Try this with the various faces of each of the boxes. Can you develop a general condition for stability?

9. You can do this qualitatively as discussed previously or take it a step further and relate the geometry of the box to the angle it can tilt at and still be stable. Can you predict the maximum angle of stability for given box dimensions?

Expected Results

The box will be stable if the center of mass (marked by the pencil) is over the base of the box in contact with the incline. Once the angle increases to the point where it is outside the base, the object will rotate.

Objects are more stable when the center of mass is closest to the incline.

Because a pizza box has a square-top face, it will be stable up to a 45-degree angle when

Figure 47-1 *The leaning tower of "pizza."*

propped up with one of the long edges placed along the incline, as shown in Figure 47-1.

If A is the length of the side of the box in contact with the incline and if B is the height of the box (for that particular arrangement) above the incline, the maximum stable angle is given by: tangent (angle) = A/B. (The angle can be found by taking the inverse tangent or *arctan*, which can be found on most scientific calculators.)

For instance, a 17-ounce box of Honey Nut Cherrios has dimensions 12 inches × 7 ¾ inches

× 2 ¾ inches. The six possible placements for this box are summarized in Table 47-1.

A few of these are illustrated in the following Figures 47-2, 47-3, and 47-4.

Why It Works

Massive objects tend to act as if all their mass was concentrated in a single point called the *center of mass*. Gravity pulling on that point causes the box to rotate about the pivot point established by the pencil. If the center of mass is above the base of support, the object tends to rotate in such a way as to remain stable on the incline. However, as the center of mass moves out from above the base, a torque is applied, which tends to rotate the object, so it rolls down the incline.

Other Things to Try

This approach can be easily extended to other shapes. You can cut a cardboard tube at an angle near the bottom edge to form a replica of the Leaning Tower of Pisa (not necessarily to scale). Try to cut it in such a way that the tube remains standing at the minimum possible angle with respect to the ground. This would make a fun challenge for a group. This can be done either by trial-and-error or by calculations based on an approximately rectangular cross-section. By the

Table 47-1

A Edge along incline (inches)	B Edge above incline (inches)	C Edge across incline (inches)	Maximum stable angle (degrees)
12	2.75	7.75	77
7.75	2.75	12	70
12	7.75	2.75	57
7.75	12	2.75	33
2.75	7.5	12	20
2.75	12	7.5	13

Figure 47-2

Figure 47-4

Figure 47-3

way, the real Leaning Tower of Pisa is currently tilted at 5.6 degrees and would topple if that angle increased 1.4 degrees, according to a tilt angle of 7 degrees (Rossella Lorenzi, Discovery Channel Online News www.discovery.com, September 1998, http://www.endex.com/gf/buildings/ltpisa/ltpnews/ltpdisc092298.htm). The mass distribution of the Leaning Tower of Pisa is not strictly that of a cylinder and currently benefits from various techniques of shoring it up.

By the same token, you can cut cardboard mailing tubes in 3", 6", 9", and 12" lengths. Then, place them on a board with a stop to keep them from sliding. As you tilt the board, each of the tubes will topple in sequence, starting with the tallest.

The Point

Objects are stable if their center of mass is within their base of support.

Energy/Momentum

Project 48
The pendulum and your physics teacher's
Ming dynasty vase.

The Idea

Energy is neither created nor destroyed; or as physicists say, energy is always conserved. This project is a test in one's confidence in this time-honored principle. This experiment can be done with any size pendulum. However, a large pendulum with a heavy mass raises the stakes and increases the suspense.

What You Need

- pendulum
  - mass: consisting of any mass that can be *securely* supported (such as a hooked mass or a bowling ball)
  - string or cable strong enough to support the mass
  - secure overhead support such as a ceiling beam that can safely handle the moving mass
- fragile object you do not want destroyed by the pendulum. You may want to start with a plastic bottle before attempting this on your more expensive pottery.

Method

1. Set up the pendulum, so it swings freely.
2. Set an object in the path of the pendulum.
3. Position the pendulum mass, so it is between the object and the equilibrium point, and *just* touching the vase.
4. *Release, but don't push* the pendulum mass from its point of contact with the object.
5. Let the pendulum go through a full excursion from where it was released, and then back.

Expected Results

If the mass was released and not pushed, it will never go higher than the point from which it was released. The pendulum will return to, but never exceed, the release point.

For people who are still in the process of developing a sense of confidence in the law of

Figure 48-1 *What goes down, must come up (to almost the same height that it was released from).*

conservation of energy, a moment of suspense may exist as the pendulum returns to its original height. In any real pendulum, there is a certain amount of friction in the point of contact and from air resistance. Because of this, the pendulum returns to a point *slightly lower* than the point from which it was released from.

Why It Works

The potential energy of an object is equal to its weight times the height it is raised to. For objects released from rest, as is the case here, there is no starting kinetic energy. So, the amount of potential energy you start with equals (or is slightly lower than) the final kinetic energy. The pendulum will never quite return to the level that it was released from because some energy is "lost to" friction as the mechanical energy is transformed into thermal energy.

Other Things to Try

A variation on this involves the related idea of *conservation of angular momentum*. If you release a pendulum in such a way that it does *not* hit the vase, no matter how many times it swings back and forth, it will *not* hit the vase.

The Point

The amount of energy contained in a moving object, such as a swinging pendulum, can neither be created nor destroyed.

Project 49
Two slopes. Different angle, same height.

The Idea

This experiment compares how much energy an object has after following several different paths. We can determine how much energy a ball has after rolling down an incline by measuring how far it rolls off a table.

What You Need

- 2 inclines supported by a ring stand or a stack of books (one incline that works well with golf balls is a vinyl bullnose section of molding available at home supply stores)
- 2 golf balls or other matched objects to roll down the incline, such as marbles, coffee cans, toy cars, or a air track glider
- meterstick or tape measure
- optional: motion sensor

Method

1. Set up the inclines at two different slopes, as shown in Figure 49-1. Allow enough space at the bottom of the incline so that the golf balls roll off the table horizontally.

2. Avoid an angle that is so severe as to cause the golf balls to bounce on the edge of the table.

3. Align the inclines so they are pointing in the same direction.

4. Hold the two golf balls at equal height above the table. This may be easier with two people.

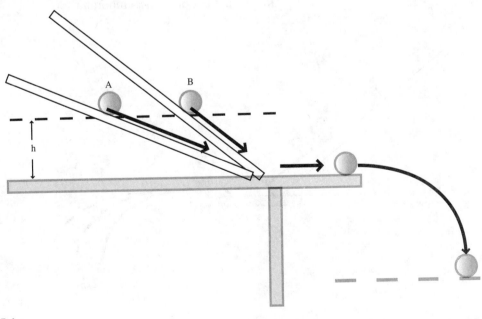

Figure 49-1

5. Predict what you think will happen with each of the balls. Which will come down with the greatest velocity? The velocity can be determined either by using a motion sensor or by comparing the point that it hits the floor after rolling off the table.

6. Release both golf balls and compare the results with your prediction.

Expected Results

Both balls should move with the *same velocity*, as they roll horizontally across the table. The balls then hit the floor at the *same distance* from the edge of the table.

Why It Works

In this, as in all other projects, energy is conserved. The energy each of the two golf balls starts off with is the same because they are released from the same height. This is equal to the object's weight times gravitational acceleration. All of this energy is converted to kinetic energy (neglecting frictional losses) when the balls get to the bottom of the incline. With *equal kinetic energy*, the objects will move at the *same velocity*.

Other Things to Try

Pick one of the slopes and hold a golf ball on the incline at each of three different places (for example at 6-inch intervals starting from the top of the incline). Predict the outcome. Release the ball from each of the three positions and compare the results with your predictions. Here, the ball starts with three different amounts of potential energy. It comes off the incline with three different velocities, as shown in Figure 49-2. According to the law of conservation of energy (neglecting friction), the potential energy (mgh) is converted to kinetic energy (½ mv²). The distance that a horizontal projectile travels is proportional to its horizontal velocity. As a result, the range or *distance* along the floor will go as the *square root* of the *height* above the table.

The Point

Total mechanical energy (consisting of kinetic and potential energy) is conserved unless some energy is consumed in overcoming friction. Objects released from the same height have equal potential energy. When this energy is converted to kinetic energy, the path the objects move toward the bottom is not important.

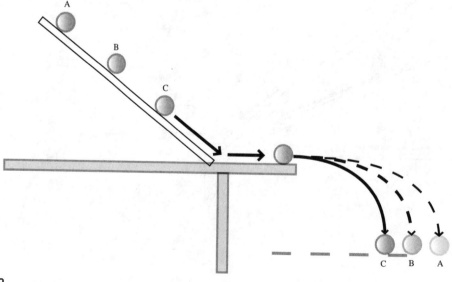

Figure 49-2

Project 50
Racing balls. The high road versus the low road. Which wins?

The Idea

Which path will take the least amount of time for a rolling ball?

- a path that is straight and horizontal, or
- a longer path that starts horizontally, dips in a curved path without excessive friction, and then returns to the same horizontal level it started from.

One path is shorter. So you might think it will take the least amount of time. Because both objects return to the same height, they wind up with the same amount of energy. How does that affect the overall time for the journey? Figure 50-1 shows the two paths.

What You Need

- 2 golf balls
- materials to build a track:
  - flexible flat wooden molding—one section 8 feet long and one section 6 feet long
  - a side board about 6 feet long
  - a couple of 2" × 4" × 6" pieces to serve as a base
  - small flat-head wood screws
  - small wooden or metal right-angle braces—1 inch corner molding will work
  - optional—a basket or plastic cup
- This type of apparatus is also commercially available, as shown in the later Figures 50-6 and 50-7.

Method

Building the track

1. Draw or sketch the shape of the curved section. This can be traced, copied, or eyeballed. A more exacting approach would be to generate a geometric cycloid and form the curve into that shape.

2. Attach the flexible track to the side board with a straight section, a downward curve returning to a second straight section. Attach the braces to the side board and secure the

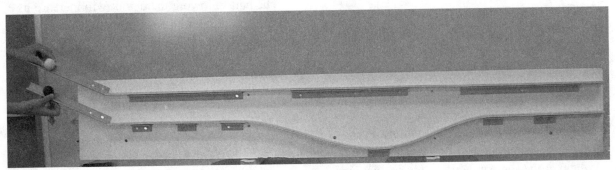

Figure 50-1 *Both ball start at the same height. Courtesy Dan Silver.*

flexible track to the braces. (Keep the profile of the screws as low as possible, so it does not interfere with the motion of the golf ball. It may be necessary to countersink the screw hole, so the screw head is below the level of the track.)

3. Attach the straight section to the side board, a few inches above the section with the detour.

4. Attach the base in such a way that the path the ball will follow is slightly tilted toward the side board. This minimizes the friction the ball encounters, but it will allow the ball to roll without falling off the track.

5. You may want to add some way to catch the balls after each race to avoid having to chase them every time.

6. A ramp of equal slope and equal length is placed at the start of each path to give objects racing down the two paths the same starting velocity. Be sure to keep any seams in the track as low profile as possible.

Racing

1. Before doing this, observers can make their prediction. Which track is fastest: a) the flat track b) the track with the detour c) both the same?

2. Release both balls from the *same height* (above the initial flat section of track) at the same time. Observe the progress of the balls. Repeat a few times to make sure the results are consistent.

Comparing energy

At the end of the track, regardless of which ball finishes before the other, measure the velocity on the final flat section. You can do this in several ways:

1. Use a motion sensor to measure the speed of the balls on each of the flat sections. If you have two motion sensors, you can measure them at the same time. If you have one, you

can do them one after the other. In either case, the most definitive conclusion will result from a good statistical sample.

2. Another way to measure velocity is to take advantage of the fact that the range of a horizontal projectile (as you saw in Project 6) depends only on its height above the ground and the velocity with which it leaves the horizontal surface. In the apparatus shown in Fig 50-1 it is clear that the starting and stopping level for each of the tracks is at a different height above the ground. This does not affect their movement relative to each other. However, it does give the track on top an advantage where it will land unless the height difference is compensated for by raising the landing level. If this is done, balls that move the same distance along the ground have the same velocity. Some designs such as a track used at Michigan State University (http://demo.pa.msu.edu/PicList.asp?DID= DID18) are built with the two tracks side-by-side, so this velocity comparison can be made more easily. Plans for a similar racing-ball track are available from the University of Maryland Physics Department at http:// www.physics.umd.edu/deptinfo/facilities/ lecdem/services/demos/demosc2/c2-11dwg .jpg.

Expected Results

The ball on the flat track will move with constant velocity (Figure 50-2).

The ball following the detour will increase its speed, so it is going fast enough to make up for the extra distance (Figure 50-3).

Once returning to the original height, the ball that went through the detour will return to its original speed. However, now it will be ahead of the ball on the flat track. The ball on the detour will reach the end of the track first (Figure 50-4).

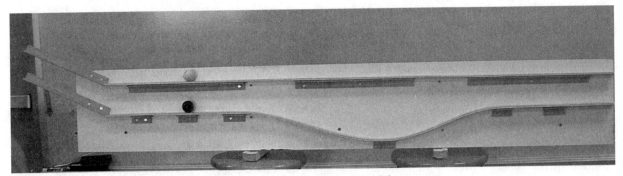

Figure 50-2 *Both balls begin with the same velocity. Courtesy Dan Silver.*

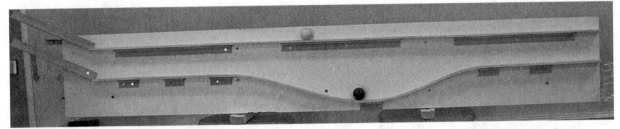

Figure 50-3 *The ball on the lower track picks up enough speed to move ahead of the top ball despite the extra distance. Courtesy Dan Silver.*

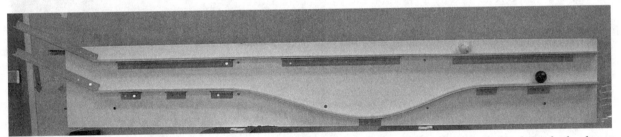

Figure 50-4 *Both ball complete their trip at the same final velocity but with the lower ball clearly in the lead. Courtesy Dan Silver.*

Why It Works

The straight path is easy. The ball travels with the same constant velocity it is given at the start. It does not gain or lose energy, except for the (relatively) small loss due to friction.

On the straight section of the curved path, the second ball travels with the same velocity as the first. As it goes downhill, it picks up speed. If the shape is right, the increase in speed will be more than enough to compensate for the longer distance.

Other Things to Try

In 1696, Johann Bernoulli challenged the most brilliant minds of his day to solve what is now known as the *brachistochrone* problem—based on the Greek "brichistos" (shortest) and "chronos" (time). Basically, the problem is this: find the path between two points at different levels that an object acted on only by gravity will travel in the *least amount of time*. This is similar to the racing ball configuration previously defined, except, in this case, the balls start from rest without any initial velocity.

Galileo previously had attempted a solution to this problem. The path Galileo defined was the circular arc connecting the two points. This, although a good approximation, was *not* the correct solution.

The correct solution was found by five mathematicians who responded to Bernoulli's challenge. This included, among others, a solution by Sir Isaac Newton, which was submitted in just one day. The path taking the shortest time was found to be a mathematical curve known as a cycloid.

A *cycloid* is defined by the equations $x = r(t - \sin t)$ and $y = r(1 - \cos t)$, where r can be thought of as the radius of the circle that sweeps out the cycloid and t is time.

An (inverted) cycloid generated (with r = 1, *t* varying from 0 to 3, and with *x* values as negative) is shown in Figure 50-5. This is actually similar to a shape generated by a pencil at the edge of circle in a Spirograph.

Although in the racing-ball scenario, we do have a slight head start in the form of an initial velocity, the cycloid curve is a good approximate minimal time path from point A to point B.

An extension to this project would be to build a track that compares a golf ball following a cycloid curve with a straight path down.

Although the large track is more fun to watch, a mini version of either of these tracks can be assembled from foamboard with a track shape cut out based on the shape in Figure 50-5 and glued to the baseboard. A clear plastic model can also work and offers the added advantage of working with an overhead projector.

A track system that can be used to study various aspects of conservation of energy can also be purchased. Two examples are shown in Figures 50-6 and 50-7.

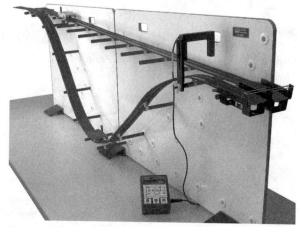

Figure 50-6 *Horizontal racing ball track. Courtesy PASCO.*

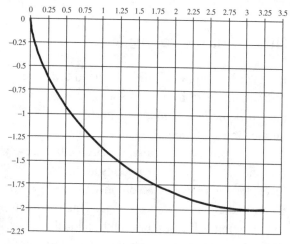

Figure 50-5 *Curve shape generated by an inverted cycloid.*

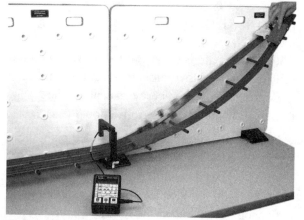

Figure 50-7 *Brachistochrone track and straight incline. Courtesy PASCO.*

The Point

The path that takes advantage of increasing the velocity during part of the trip takes less time than one of constant velocity.

This project demonstrates that (aside from frictional losses) energy is conserved. As the potential energy is reduced, it is transformed into kinetic energy. Because both balls have returned to their original height and finished with the same velocity, we confirm that kinetic energy is conserved, regardless of the path traveled.

Project 51
Linear momentum. Where can you find
a perfect 90-degree angle in nature?

The Idea

Can you think of anything that forms a perfect right angle in nature? One example of a right angle is the fracture plane of body-centered cubic crystal, such as calcite. This experiment explores another example of a natural right angle that results from an elastic collision between two objects of equal mass.

What You Need

- 2 low-friction objects of equal mass. Hover Pucks are excellent for this. Pennies on a smooth table can also work.

- flat, level surface

- protractor

- 2 lengths of string

Method

1. Mark the starting point of the first (stationary) puck. (Find a reasonable flat place on the floor to prevent the first puck from drifting away prematurely.)

2. Place the second puck a short distance from the first one. (Do what works for you, but 18 inches may be a good starting point. If you are doing this with pennies, you probably want to shorten this to a few inches.)

3. Push the second puck (the shooter) toward the stationary puck. Aim so the collision is at a glancing angle, rather than head on, hitting the stationary puck off-center as shown in Figure 51-1.

4. Mark one or more points along the path of each of the pucks after the collision. You may find a few extra sets of hands are helpful here.

5. Take a piece of string and place one end at the center of the stationary puck. Place the other along the path it traveled.

6. Place the other piece of string with one end also at the center of the stationary puck and the other along the path of the puck.

7. Measure the angle between the two strings.

Figure 51-1 *A puck (moving toward you) about to hit a stationary puck.*

8. Get a better statistical sample by repeating this a few times and taking the average. If your collision is too direct, your angle will be zero or close to it and shouldn't be included in your average.

Expected Results

The angle the two paths make should form a right triangle, as picture in Figure 51-2. One exception is if the moving puck hits along the centerline of the stationary puck, it may stop and send the stationary puck moving along the same path.

Why It Works

When two objects collide, momentum (given by mass times volume) is conserved. However, for elastic collisions, such as are being explored here, kinetic energy is *also* conserved. The only way kinetic energy can be conserved is for the colliding objects to form a perfect right triangle.

Kinetic energy is given by one-half the mass times the velocity squared (or $\frac{1}{2}mv^2$). If v_c is the velocity of the shooter before the collision, and v_b is the velocity of the shooter after the collision, then v_a is the velocity of the stationary object after the collision. (There is, of course, no velocity for the stationary object before the collision because it is stationary.) Conservation of energy gives us:

$$\tfrac{1}{2}\,mv_a^{\,2} = \tfrac{1}{2}\,mv_b^{\,2} + \tfrac{1}{2}\,mv_c^{\,2}$$

Because the mass is the same for each object, this reduces to:

$$v_c^{\,2} = v_a^{\,2} + v_b^{\,2}$$

This is the format of the familiar *Pythagorean formula* ($c^2 = a^2 + b^2$), which applies *only to right triangles*. Figure 51-3 may help visualize this. Because the velocities must be consistent with this condition, the angle the two Hover Pucks move at must be a right angle.

Other Things to Try

This experiment can also be done on a pool table. However, I have yet to know of anyone actually improving their game by applying the laws of physics. The felt of a pool table may introduce enough friction to prevent the collisions from being completely elastic. A sheet of foamboard, as shown in Figures 51-4 and 51-5, can help the collision be sufficiently elastic to be at (nearly) a 90-degree angle.

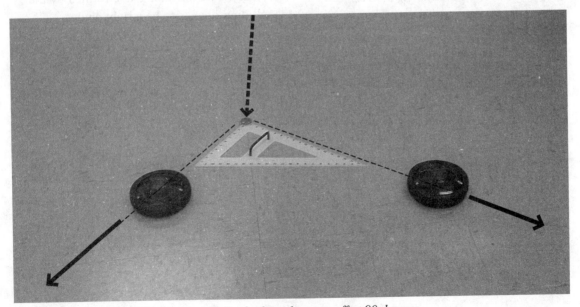

Figure 51-2 *After the (nearly elastic) collision both pucks move off at 90 degrees.*

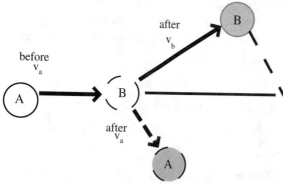

Figure 51-3 *Conservation of energy requires that the Hover Pucks move off at right angles.*

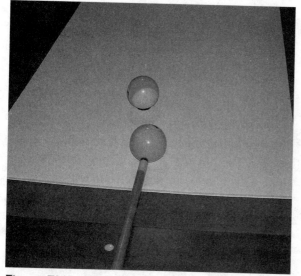

Figure 51-4 *The solid ball is hitting the stripped ball.*

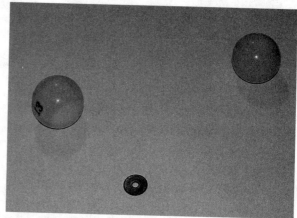

Figure 51-5 *Elastic collision from objects of equal mass.*

When researching the collisions of subatomic particles, sometimes the incoming particle strikes a stationary particle of the same mass. A collision between a moving proton and a stationary proton meets these criteria. Because subatomic particles follow the law of conservation of momentum, the trajectory of the two equal mass particles (absent any magnetic fields) as they move off is at a characteristic 90-degree angle.

The Point

Conservation of kinetic energy for elastic collisions requires the angle formed by the colliding objects to be a right angle.

Project 52
Elastic collisions.

The Idea

When one object strikes another in such a way that the objects bounce off each other, the collision is said to be elastic. When this happens, whatever momentum you start off with, you have at the end. In the case of an elastic collision, the objects *also* move off with the same overall kinetic energy they started with. In this project, we explore what happens when collisions are elastic.

What You Need

- 4 pool balls (or hardballs or golf balls)
- track for the balls to roll in one dimension (This can easily be set up by taping 2 meters sticks to a smooth board)
- large ball, such as a basketball
- smaller ball, such as a ping-pong ball
- optional—a Newton's cradle, as shown in Figure 52-1

Method

One ball hitting three

1. Place three balls of equal mass in the track.
2. Place the fourth ball a few inches away in the track.
3. Roll the ball, so it collides with the other three.

(There are several other ways to do this, including a Newton's cradle or four equal mass sliders in an air track.)

Big ball/small ball

1. Place the small ball on top of the large ball.
2. Drop both balls together. (Caution: do this in a place where, if the small ball goes flying off, it won't break anything and won't hurt anyone. If the balls you are using are small enough, you may be able to do this in a clear plastic vertical guide or in a large graduated cylinder.)

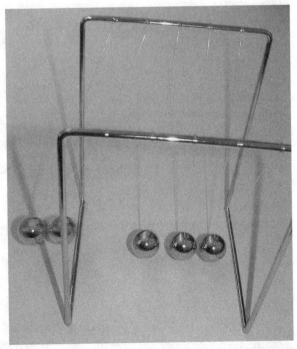

Figure 52-1 *Conservation of momentum and kinetic energy requires that two balls hitting the group always causes two balls to slide out.*

Expected Results

One ball hitting four

The incoming ball comes to a dead stop, as shown in Figures 52-2 and 52-3. The outermost stationary ball moves in the same direction and at the same velocity as the incoming ball. The other three stationary balls do not move.

Big ball/small ball

The balls bounce together. After striking the ground, the smaller ball flies off with much greater velocity than the large ball.

Why It Works

With the stack of balls, it is not hard to understand how the momentum of the incoming ball is transferred to the ball that gets knocked out of the stack. This is a clear illustration of conservation of linear momentum.

But *why is only one ball knocked out* of the stack? Why, for instance, do we never have two balls knocked out with each taking one half of the momentum of the incoming ball? That would also be perfectly consistent with the law of conservation of momentum. The problem is these collisions are elastic collisions, which means not only is momentum conserved, but kinetic energy is also conserved. The only way this can happen is for a single ball to emerge from the stack with the same momentum as the incoming ball.

With the large and small balls, the large ball having a larger mass conserves momentum by causing the smaller ball with a lower mass to fly off with a larger velocity.

Other Things to Try

A Newton's cradle, as shown in the previous Figure 52-1, is another good way to study elastic collisions. In a *Newton's cradle*, two balls never rebound when struck by a single ball and three balls never rebound when struck by two balls.

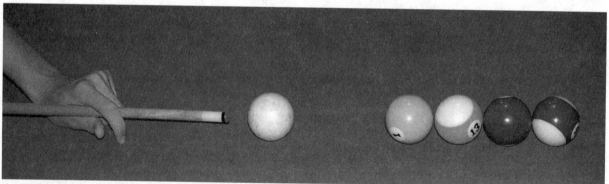

Figure 52-2 *One ball hitting the group—before collision. Courtesy Dan Silver.*

Figure 52-3 *After collision—results in only one ball knocked out. Courtesy Dan Silver.*

This is the result of both conservation of linear momentum and conservation of energy.

The Point

In an *elastic collision*, both linear momentum and kinetic energy are conserved.

When momentum is transferred from one object to another, a larger velocity compensates for a smaller mass. In the case of an elastic collision between objects of equal mass, this condition can be met only when the same number of balls move after the collision as were moving before.

Project 53
Inelastic collision. Sticking together.

The Idea

When objects collide, they either bounce off each other or they stick together.

What You Need

- 2 low-friction carts
- Velcro or duct tape
- low-friction track (optional)
- motion sensor
- index card
- tape

Method

1. Measure the mass of each of the two carts.

2. Attach Velcro to the end of each of the carts, so when they meet, they stick together. You could also use duct tape formed into a loop and attached sticky-side out to each of the carts.

3. Set up the motion sensor at one end of the table.

4. Place the first cart near the motion sensor. The Velcro side should be in front, away from the motion sensor. If you have a low-friction track, place the cart on the track in a line pointing away from the motion sensor.

5. Place the second cart near the midpoint of the table with the Velcro in the rear.

6. It may be helpful to attach an index card to the back of the first cart to make it easier for the motion sensor to pick it up. If you can get away without doing this, you can avoid air resistance that could slightly affect your result.

7. Set up the motion sensor to read distance and velocity versus time.

8. Start the motion sensor. It should be on the cart setting and focused on the card of the first cart.

9. Give the first cart a push in the direction of the second cart. It should be slow enough to get a good reading from the motion sensor, but fast enough to rear-end the second cart and push it along for at least a few seconds or more. The first cart collides with the second totally *inelastically*, which means they stick together after the collision.

10. When both carts stop moving, stop collecting data from the motion sensor.

11. From the motion-sensor graphs, find the velocity of the first cart before the collision and the velocity of both carts joined together after the collision. The graph of velocity versus time may be a little erratic right after the collision, reflecting the impact. Pick a point where the velocity has settled down.

12. Momentum is defined as mass times velocity. Compare the momentum before and after the collision.

13. Kinetic energy is defined as ½ times the mass times the velocity squared. Compare the kinetic energy before and the kinetic energy after the collision.

Figure 53-1 *Inelastic collision with one or both carts initially moving. Courtesy PASCO.*

The experimental setup is shown in Figure 53-1. (This actually shows a motion sensor at *both* ends. The previous procedure uses only one motion sensor, but this can easily extended to include both carts in motion. For simplicity, we will start out with one of the carts stationary.)

Expected Results

The momentum of *both carts before* the collision should equal the momentum of *both carts after* the collision.

Before the collision, one of the carts is stationary, which means it has no momentum, so the moving cart is the only one with momentum before the collision.

After the collision, both carts stick together and move off with the same velocity. The combined mass of the two carts together times their combined velocity is the momentum after the collision.

Figure 53-2 shows the position versus the time graph before and after the collision obtained by a motion sensor. Notice how the slope of the line abruptly drops, indicating the collision.

Figure 53-3 shows the velocity versus time graph before and after the collision. The velocity before and after can be determined directly from the graph. You can notice a slight downward slope indicating some slowing of the carts due to friction. This is not a showstopper for the experiment, but it shows the extent to which an air track can improve the overall results.

The most reliable velocity measurement is immediately before the collision. The collision shows some bouncing around and variability in the velocity for a short period until the two carts stick together and move as one. This provides some insight into the nature of inelastic collisions, which result in the loss of kinetic energy (but not linear momentum). The most reliable postcollision velocity to use is the point where a new horizontal line begins. The results should be fairly accurate, but some losses due to friction may be encountered without an air track. Also, excessive mass can load down the wheel bearing and increase the losses to friction.

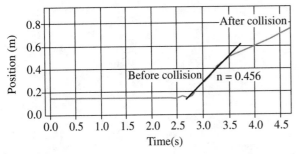

Figure 53-2 *Motion sensor measurement of distance versus time for an inelastic collision between a moving car and a stationary cart.*

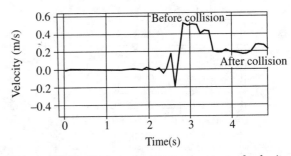

Figure 53-3 *Motion sensor measurement of velocity versus time for an inelastic collision between a moving car and a stationary cart.*

Why It Works

The total momentum before an inelastic collision equals the total momentum after.

However, unlike an elastic collision, the kinetic energy for an inelastic collision is less after the collision.

Other Things to Try

1. If you have two motion sensors, you can repeat this with both carts initially in motion. You can get the velocity of each cart right before the carts collide and the velocity of both carts together after the collision. Both sensors will read a positive velocity before the collision as the distance from the sensor increases. After the collision, one sensor reads a positive velocity, while the other reads a negative velocity. Results are shown in Figure 53-4.

2. Compare elastic and inelastic collisions using so-called "happy/sad" balls. The balls appear to be completely identical. However, one is elastic and bounces back from the floor, while the other is inelastic and doesn't bounce at all.

3. Hang elastic and inelastic balls to form a pendulum. Stand up a wooden block in front of the pendulum. Swing the inelastic ball first, so it doesn't quite knock the block over. Compare that to what happens with the elastic ball swung under the same conditions. Elastic collisions result in *double* the change in momentum as an inelastic collision under the same conditions. This is because in the elastic collision, the momentum not only stops (as it does in the case of an inelastic collision), but it also reverses itself in the other direction.

The Point

In an elastic collision, the objects bounce off each other in such as way that linear momentum and kinetic energy are conserved. This is true to the extent that no external force occurs during the collision.

In an inelastic collision, the objects interact in such a way that linear momentum is conserved (as long as no forces affect the collision). However, kinetic energy is not conserved in an inelastic collision. In a perfectly inelastic collision, the objects stick together and move after the collision as if they were a single object.

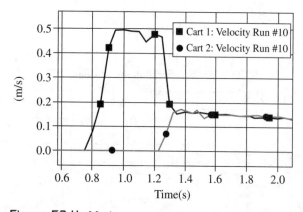

Figure 53-4 *Motion sensor measurement of velocity versus time for an inelastic collision between a moving cart and a second moving cart.*

Project 54
Impulse and momentum. Eggstreme physics.

The Idea

What do you think will happen if you throw a raw egg as hard as you can at a blanket held vertically? There is really only one way to find out. This experiment gives you an opportunity to explore the relationship between momentum and impulse.

What You Need

- 1 raw egg
- blanket
- 3 people
- mop and paper towels for cleanup (very optional)

Method

1. Hold the blanket vertically, with the bottom edge curled out to form an overhang, as

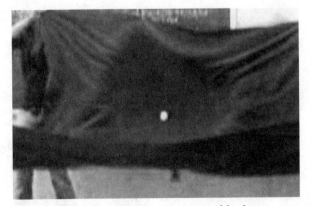

Figure 54-1 *Throwing a raw egg at a blanket.*

shown in Figure 54-1. This requires at least two people.

2. The third person throws the egg at the blanket. Don't hold back. Give it a good shot. You can throw as hard as you can without having the egg break in your hands as you throw.

Expected Results

You know what will happen if you throw a raw egg against a cinder-block wall. However, if the egg is stopped by the blanket, the deceleration occurs over a sufficiently long time, which prevents the egg from breaking.

Why It Works

Momentum is changed by a force exerted over time. The ability to change an object's momentum is called *impulse*, which is defined as the force exerted multiplied by the time. Anytime an object is brought to rest, the change in momentum equals the momentum the object had to start, applied in the opposite direction. The impulse to bring that object to rest can come from any combination of force and time, which when multiplied, equal the momentum change.

If the object's momentum changes in a short time, which would occur if an egg is thrown at a cinder block wall, the force is greater. However, if the egg is thrown at a blanket that brings the egg to a stop over a much longer period of time, the force is much smaller. This is why the egg does not break when it's throw, at a blanket.

Other Things to Try

Check out the ESPN video *Relaxing with Momentum* that shows what happens if you drop a watermelon from a diving board onto concrete, compared with dropping a watermelon into the water.

The Point

The momentum to stop an egg thrown against a wall and a blanket is the *same*. However, the force in the case of the blanket is spread over a greater time and is much smaller.

Project 55
Using gravity to move a car.

The Idea

Energy can neither be created nor destroyed. One form of energy, however, can be changed into another form of energy. For instance, the combustion that occurs in a car engine produces heat energy, which, in turn, is converted to mechanical motion by the motor. In this project, you convert gravitational potential energy into mechanical energy.

What You Need

- wheels: old CDs or DVDs
- post to hold the weight: a dowel 1 m in length and ½ to 1 inch in diameter
- 1 kg weight
- string
- axles (dowels work well)
- materials to build the body of the car
- plumbing fitting or a block of wood to attach the post to the car body
- tape measure
- stopwatch

Method

1. Assemble the car. Use Figure 55-1 as a guide, but feel free to develop and build your own concept.

2. The basic design criteria for the car includes:
   - The CD (or equivalent) used for wheels should turn freely.
   - The descending mass should freely turn an axle to produce forward motion of the car.
   - The car should be balanced with the mass in both elevated and falling positions. (Remember, a 1 kg mass raised 1 meter above the ground can exert a lot of torque that could topple the car.)
   - The mass should fall *onto* the car after it descends (rather than dragging along the ground, which can limit how far it goes).
   - There should be enough symmetry between left and right, so the car moves forward, rather than turning.

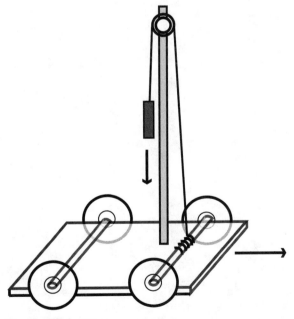

Figure 55-1 *Kilogram car.*

3. Wrap the string around one of the axles and attach the other end to the weight. Decide if you want front-wheel or rear-wheel drive and wrap accordingly.

4. "Arm" the car by raising the mass and leaving some of string still wrapped around the axle it will turn.

5. Orient the car on a designated course, and then release the mass.

6. Measure how far it goes.

7. If several individuals or groups are involved, it may be fun to do this as a race to see which car goes the farthest or which car crosses the finish line first.

Expected Results

It might take a few hours to build the car(s), depending on what materials are available. If built correctly, the car should move as the mass descends. Once the mass falls and comes to rest (somewhere on the car), the car has enough momentum to keep moving.

Why It Works

A mass at a height contributes an amount of energy equal to its mass times the height above the ground times gravitational acceleration. In this case, 1 kg dropping a distance of 1 meter will contribute 9.8 joules of energy. (One joule is about the amount of energy needed to lift an apple 1 meter.)

Other Things to Try

A variation of this is to use a mousetrap to supply the energy. The mousetrap is started in the open position. As with the previous kilogram car, the potential energy stored in the spring of the mousetrap is transferred to the forward motion of the car.

The Point

Energy is conserved. The potential energy you start with equals the kinetic energy given to the car (plus any energy lost to friction).

Project 56
How can CSI measure muzzle velocity?
The ballistic pendulum.

The Idea

You can tell how fast an object, such as a bullet, is moving by how it affects the momentum of another object.

What You Need

- projectile and launcher:
  - golf ball and a ramp
  - precision projectile launcher
  - hand-thrown hardball or golf ball
- balance to measure mass
- box
- tape
- string
- ring stand or improvised support
- protractor
- meterstick

Method

1. Build (or buy) a receiver box that meets the following conditions:
   - The moving projectile should be caught and retained in the box. Lining the interior with double-sided duct tape or foam rubber can serve to capture the projectile.
   - The box is suspended from the ring stand.

2. Suspend the box from the ring stand using the string.

- Determine the mass of the projectile.
- Determine the mass of the catcher box and include any of its contents.
- Direct the projectile toward the catcher box and allow the box with the projectile caught inside to swing upward.
- Measure the maximum height above the initial starting position that the catcher box and the projectile reach. (Alternatively, the angle and radius of the pendulum formed by the supported catcher box can be recorded.)

The basic design of the apparatus is shown in Figure 56-1.

The velocity of the projectile can be determined from the following equation:

$$v = \left(\frac{m + M}{m}\right) \sqrt{2gh}$$

where m is the mass of the projectile, M is the mass of the "catcher box," g is the gravitational constant $= 9.8$ m/s^2, and h is the height that the

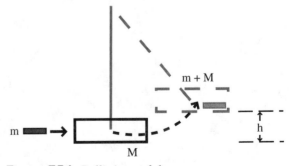

Figure 56-1 *Ballistic pendulum.*

catcher box is lifted by the momentum of the projectile.

Expected Results

The more momentum the projectile has, the greater the angle the pendulum swings through. For a given mass, a greater velocity causes a greater projectile.

Why It Works

The momentum of the projectile ("bullet") is transferred to the box. The greater the momentum of the projectile, the higher the box is driven.

Other Things to Try

A semiqualitative version of this consists of placing a cardboard box filled with tissue paper on the floor and comparing the distance it moves with balls rolled or thrown into it at different speeds. This can be refined by measuring the coefficient of kinetic friction between the box and the floor and the distance the box slides along the floor. Use of the equations of motion can be solved to determine the initial velocity of the box.

Various ballistic pendulum designs are also available from science supply companies.

The Point

The velocity of a moving object can be determined by measuring its effect after colliding with another object of known mass. The *ballistic pendulum* is based on conservation of momentum applied to a pendulum. The larger the velocity of the incoming object, the higher the pendulum swings.

Project 57
Angular momentum. Riding a bike.

The Idea

A bicycle is unstable when it is stationary. If you try to balance a bike that is not moving, it will fall. This experiment explores how it is possible to defy gravity and ride a bike.

What You Need

- rope—about a meter (about a yard) in length
- bicycle tire
- dowel or section of broom handle to fit snuggly into the axle of the tire

Method

1. Attach the rope to the axle of the bicycle tire. The tire should be able to turn freely around the attachment point. If there is no axle, insert a cylindrical piece of metal or a dowel for the tire to rotate around.

2. Suspend the tire from the rope, so it tangles freely. Either hold the other end of the rope in your hand or attach it to something overhead.

3. Holding the tire vertically by the axle, spin the tire.

4. With the tire spinning, release the tire, so it is supported by the rope.

5. Try this with the tire spinning rapidly and slowly, as shown in Figure 57-1.

Expected Results

The first thing to notice is the tire will remain in close to the vertical position. Also, it doesn't take much of a spin to keep the tire stable. You should also be able to observe that if left alone with the tire near the vertical position, the spinning tire rotates about the pivot point established by the rope. This is called *precession*.

Figure 57-1 *It is easy for a spinning bicycle wheel to remain vertical. Courtesy PASCO.*

Why It Works

A full explanation of this simple situation can get complicated in a hurry. The basic idea is that a spinning tire has angular momentum. Gravity tries to rotate the tire from the vertical position that it is spinning in to the horizontal position that a nonrotating tire would be in. The force exerted by gravity produces a torque that is at *right angles* to *both* the force exerted by gravity and the direction of the angular momentum, which is along the line of the axle. This not only keeps the wheel from falling, but it also causes it to precess in a circle. This forms the basis for gyroscopic movement.

Other Things to Try

Like a bicycle, a toy gyroscope becomes stable only when it has sufficient angular momentum to counterbalance the pull of gravity. As an extension, study a gyroscope. Observe what happens when its turning axis is displaced from a stable position. What affects the rate of precession?

The Point

A bicycle tire is stable when it is rotating because the tire has angular momentum. The gravitational attraction of the Earth exerts a force that would pull the tire to a horizontal position were it not for the angular momentum of the spinning tire. The interaction of the torque caused by gravity and the angular momentum of the spinning tire results in the precession of the tire about the pivot point.

Project 58
Moment of inertia. Ice skaters and dumbbells.

The Idea

How do ice skaters get spinning so rapidly? Where do they suddenly get the energy? Are they violating conservation of energy? This project explores how this works. The term "dumbbells" should in no way be construed to refer to the skaters or the experimenters (or the writer of this book for that matter). They refer to actual dumbbells.

What You Need

- (low-friction) rotating stool
- 2 masses, such as a pair of dumbbells (5 kg or greater)
- 1 person
- bicycle tire mounted on an axle

Method

1. Sit on the stool while holding the two masses.
2. While sitting and balancing on the stool, rotate by pushing off with your feet or by being pushed by someone else.
3. Lift both feet from the floor.
4. Start with both masses extended out at arm's length. Then, bring them in close to your body, as shown in Figure 58-1.

Expected Results

Extending your arms slows you down; bringing them closer to your body lets you speed up. The rotational speed (or angular velocity) is greater when the masses are closest to the center of

Figure 58-1 *Pulling in your arms causes you to spin faster. Courtesy PASCO.*

149

rotation. This works best if the stool rotates very freely with a minimum of friction. The stool can be picked up at a yard sale or purchased commercially. There is also usually less friction for a lower-mass person.

Why It Works

This project is an illustration of the principle of conservation of angular momentum. Angular momentum for a rotating mass increases either if the object rotates faster or if there is more mass at a greater distance from the center of rotation. If the mass is moved further from the center of rotation, the angular velocity must increase to keep the angular momentum constant.

Other Things to Try

Spinning a bucket on a rope

1. Sit on the stool with both feet off the floor.

2. Swing the bucket in a circular path parallel to the floor.

3. Observe the effect of swinging faster and slower.

4. Observe the effect of using longer or shorter lengths of rope.

The main thing you notice is you rotate in the *opposite* direction that the bucket is swung. The faster the bucket rotates in a clockwise direction, the faster you rotate in the counterclockwise direction. A longer section of rope turns you faster than a shorter section.

The Point

A spinning object, such as a skater, must conserve angular momentum. A movement of mass closer to the center of rotation is compensated by an increase in how fast the skater rotates.

Project 59
What caused Voyager to point in the wrong direction?

The Idea

The Voyager program produced some of the most remarkable spacecraft ever built, providing an unprecedented view of nearly every planet in the solar system. As Voyager II approached the planet Uranus, the spacecraft precisely trained its instruments on that planet's surface to take advantage of the short window of opportunity to get high-resolution, close-up pictures. Just as the reel-to-reel tape recorder on the spacecraft turned on to capture this historic moment, the orientation of the spacecraft was thrown out of whack and the navigational system had to compensate with last-minute corrections. This project helps you investigate why this could have happened.

What You Need

- freely rotating stool
- bicycle tire with handles (the more massive the better)
- 2 people

Method:

1. Sit on the rotating stool as before.

2. Have someone hand you the bicycle tire that had previously been set into motion. See Figure 59-1.

3. Start with the tire in a horizontal position.

4. With the tire rotating, turn the tire upside down (so it is now spinning in the opposite direction).

5. Wait a few seconds. Then, turn the tire upside down again.

Expected Results

You will rotate on the stool in the opposite direction from which the tire is rotating. While changing the position of the tire, you will feel a surprisingly strong force, as if you were pushing against a solid wall (Figure 59-2).

Why It Works

Starting from rest, both the person and the stool have zero angular momentum. For angular momentum to be conserved, it is necessary that the person on the stool rotate in the opposite direction as the rotation of the tire to preserve angular momentum. When the reel-to-reel tape recorder on Voyager II turned on to record the spectacular images of Uranus for the first time in history, it began turning with a new angular momentum. See Figure 59-3. Just like the person on the stool, Voyager began to turn in the

Figure 59-1 *A rotating wheel has angular momentum. Courtesy PASCO.*

Figure 59-2 *Angular momentum is conserved. Courtesy PASCO.*

Figure 59-3 *Conservation of angular momentum required a positioning correction aboard Voyager II to compensate for a rotating reel-to-reel tape recorder. Source NASA.*

opposite direction as the tape recorder to conserve angular momentum.

Other Things to Try

This principle is demonstrated by a toy train running on a circular track. The track is mounted on a platform attached to a freely rotating support. As the train moves in one direction, the platform rotates in the opposite direction to conserve angular momentum.

The Point

Angular momentum is conserved. This applies to the case where a system starts with zero angular momentum.

Project 60
Moment of inertia. The great soup can race
or that's how I roll.

The Idea

If two identical cans are released from rest to roll down an incline at the same time, will the top can catch up with the bottom can? Does it matter what the contents of the cans are? This project deals with how things roll.

What You Need

- 2 cans of thick soup (such as mushroom soup)
- 1 can of thin soup (such as chicken broth)
- incline

Method

Part 1

1. Verify that the external shape of each of the three cans is the same.

2. Set up an incline about 1 meter in length. The height should be about 30 cm.

3. Hold the two cans of mushroom soup on the incline with a distance of about 10 cm between them, as shown in Figure 60-1. Call the top can A and the bottom can B.

4. Predict what will happen when the cans are released: a) the distance between them will increase b) the distance between them will decrease c) the distance between them will remain the same.

Part 2

1. Place one of the mushroom soup cans (A) and the can of broth (C) on the incline with A 10 cm higher than C.

2. Predict what will happen when the cans are released: same options as number 4.

3. Try this again with can C as the upper can this time.

Expected Results

The two cans with the same contents will accelerate at the same rate. Because both cans start from rest, the distance between them remains constant. However, the mushroom soup has a greater density than the broth and it will roll more slowly.

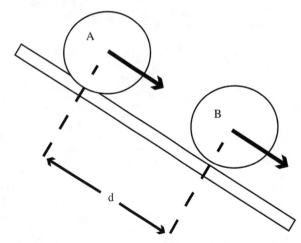

Figure 60-1 *How will the distance, d, change as the can roll down the slope?*

Why It Works

When an object rolls, some of its energy is associated with moving from one point to another (called *translation*). The more mass the object has and the faster it goes, the more energy it has. In addition, some of the energy of a rolling object is related to the fact that it is rolling. The amount of this rotational energy is related to the way the mass is distributed around the center of rotation. A greater density center (mushroom soup) requires more energy to roll at the same rate as a lower density center (broth).

Physics alert: A property called *moment of inertia* measures how mass is distributed around a center of rotation. A can of dense soup has a greater moment of inertia than a can of thin soup and, as a result, ties up more energy as it rolls.

Other Things to Try

An interesting follow-up would be to drop both types of soup cans from a distance and reestablish the principle that they both *accelerate* at the same rate in free-fall. Translation (falling) is different than rolling. The effect of rolling makes the difference here and gives the less-dense soup the advantage.

The Point

A force applied to a cylindrical object can cause it either to translate or rotate or some combination of both.

Project 61
Making waves. I thought I node this.

The Idea

Many of the things physics deals with are waves. This includes sound, light, and vibrations in matter. It is helpful to use vibrating objects, such as we do in this project, to help visualize more abstract waves, such as electromagnetic waves, which include light.

What You Need

- slinky
- coiled spring "snakey"
- string
- stopwatch
- tape measure
- short, thin, metal pole or wooden dowel (10 cm in length and 2 mm in diameter should work well)

Method

For the following, be careful when working with stretched springs. Be careful not to let the spring go accidentally, which could cause the spring to whip around and possibly hit someone.

Longitudinal wave with a slinky

1. Stretch the slinky to about double or triple its original length. This requires two people.
2. Measure the distance between the two ends of the slinky.
3. From one of the ends, pull back on the slinky in the direction that the slinky is stretched by a few inches and release.
4. Observe the pulse moving from one end of the slinky to the other. Time how long it takes to go the measured distance from one end to the other.
5. Calculate the velocity of the pulse by dividing the distance the wave travels by the time it takes. (Use consistent units, meaning if you measure the distance in meters, the velocity will be in meters per second. If you measure the distance in inches, the velocity will be in inches per second.)
6. Increase the tension and calculate the velocity.
7. Decrease the tension and calculate the velocity.

Transverse wave with a coiled spring

1. Stretch the snakey (coil spring) to about double or triple its original length.
2. Measure the distance between the two ends of the spring.
3. From one of the ends, displace the coil along the floor perpendicular to its length by a few inches and release.
4. Measure the velocity of the pulse with increased and decreased tension as in the previous section.

Reflection from fixed and unconstrained ends

1. Working with the coiled spring, release a transverse pulse down the spring.

2. Both ends of the spring should be held tight.

3. Observe what happens to the pulse when it reaches the end held tightly in place.

4. Now, insert the dowel or a metal rod through one of the ends of the coil. The dowel should pass through one or a few coils in such a way that the coil is able to slide freely on the dowel.

5. Again release a transverse pulse down the spring and observe what happens when the pulse reaches the end.

Waves crossing into different media

1. Connect the slinky and the coiled spring together with a string.

2. Observe what happens to a pulse sent from the slinky side.

3. Now see what happens to a pulse sent from the coiled spring side.

4. In which section (slinky or coiled spring) does the pulse move fastest? If your springs are long enough, time it and calculate the velocity. With everything else equal, a less-tense spring gives you a little more time to make the measurement.

Superposition—constructive and destructive interference

1. Place a coil on the floor. Hold the coil from both ends using two people—one on either side. Stretch it fairly tight.

2. Each person holding the coil simultaneously should release a pulse the same size from the same side of the coil. Observe what happens.

3. People holding the coil should now release a pulse the same size *from the opposite sides* of the coil. How does this compare with the pulse released from the same side?

Standing waves and nodes

1. Place a coil on the floor. Hold the coil from both ends using two people—one on either side. Apply moderate tension.

2. One person should hold the end of the coil stationary.

3. The other person should begin shaking the coil, slowly at first.

4. Observe what happens, for a given tension, as you increase the frequency of the vibrations. A *node* is a point on the wave that does not move while the wave vibrates. See Figure 61-1.

5. Quantify this by counting the number of nodes as a function of the frequency of the vibration. The frequency can be determined by the number of seconds it takes for ten vibrations (back and forth) divided by ten.

6. Observe what happens as you increase the tension in the coil.

Expected Results

Longitudinal wave

The velocity of a longitudinal wave increases as the tension increases, but is not dependent on the amplitude.

Figure 61-1 *Transverse wave.*

Transverse wave

Similarly, the velocity of a traveling wave increases as tension increases.

Reflection

When a wave comes to the end of the spring that is rigidly held, the wave reflects by flipping over to the opposite side. If the spring is free to move, the wave reflects, but begins its return on the same side it came from.

Different media

When a wave goes from one spring to another, the wave is partially reflected and partially transmitted into the second spring.

Superposition

Pulses coming from the same side of the coil add together to form a larger combined pulse. This is called *constructive interference*.

Pulses coming from the opposite side of the coil negate each other, resulting in a smaller pulse at the point or even no pulse. This is called *destructive interference*.

After combining, the original pulses continue moving through the spring.

Standing waves

The greater the tension and the more rapidly the spring is shaken, the greater the number of nodes formed.

Why It Works

Waves exhibit characteristic properties that include: traveling waves, standing waves, reflection, moving between different media, superposition, and interference.

Other Things to Try

1. Standing waves can also be shown by vibrating (or rotating) a string held under tension by two vertical supports. Waves can be generated by a motor or a speaker, as picture in Figure 61-2.

2. A fun way to do some of these investigations is to use glow-in-the-dark rope available from PASCO.

3. A significant, but often overlooked, point is the connection between the frequency of the traveling waves and standing waves in a coil. They should be the same, and the previously techniques developed are a good way to verify this. A *standing wave* is really a mixture of many standing waves *traveling* back and forth along the coil. The overall standing wave pattern is generated by the traveling waves interfering constructively and destructively. The relationship between standing waves and traveling waves is used when measuring the speed of sound by determining the wavelength of a resonant *standing* wave. This works because the speed of both is the same.

The Point

The properties of waves, including movement in a medium and reflection, can be observed in springs.

Figure 61-2 *Generating a standing wave. Courtesy PASCO.*

Project 62
Rolling uphill.

The Idea

Most people will not see the results of this little demonstration coming, especially if it's done as an immediate follow-up to the previous project. You can make a can actually roll uphill for a short distance without violating any of Newton's laws.

What You Need

- coffee can
- weight—an old battery should work well
- strong glue—such as Gorilla glue or epoxy
- incline

Method

1. Glue the weight to the inside of the coffee can.

2. Conceal the interior with the plastic cover of the coffee can.

3. Place the can on the incline with the mass on the uphill side of the can's centerline. See Figure 62-1.

4. Ask someone observing this what they think will happen. You can add to the effect by creating the illusion that you are investigating one of the situations of the previous discovery by using two coffee cans with weights.

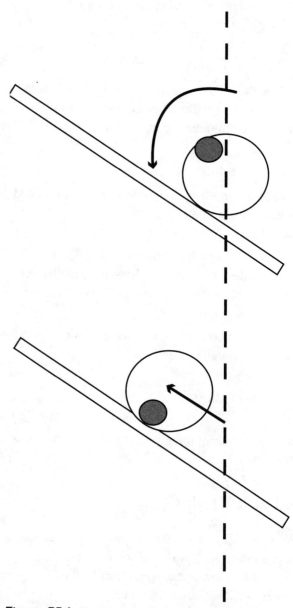

Figure 62-1 *Gravity creates a torque on the can causing it to roll uphill.*

Expected Results

Once released, the can rolls uphill a few inches.

Why It Works

Gravity pulls on the weight, which exerts a torque on the can. If the weight is positioned on the uphill side of the can's centerline, the can will roll uphill (until the weight is brought directly under the can's center of gravity).

Other Things to Try

Another similar idea is to attach a rubber band with a weight in the middle to the inside of the can. As the can rolls, the rubber band is wound up. When the can rolls, it reaches a turning point, and then returns back in the direction from which it came as the rubber band with the weight unwinds.

The Point

If the force exerted by gravity on an object's center-of-mass produces a torque on the object, the object can briefly roll uphill.

Project 63
Getting around the loop. From how far above the ground does the roller coaster need to start?

The Idea

Energy is conserved. An object has potential energy because it is a certain distance above the ground. As it rolls down an incline, some of its energy changes into kinetic energy

What You Need

- aluminum track used to secure shelving in bookcases. This is the type hung vertically on the sides of the bookshelf and holds clips that support the shelves. This track should be flexible enough so you can bend it to form a circular shape. (You don't want the heavy-duty shelving tracks you would use to set up a bookshelf to hold all your physics textbooks.)

- 1 marble or steel ball that smoothly rolls in the track

- circular object to use a guide, such as a small bucket or a one gallon paint can

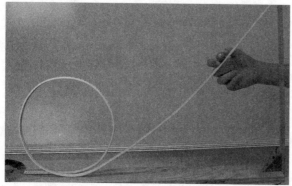

Figure 63-1 *Where must the ball be positioned to make it around the loop without falling?*

- frame to mount the frame on
    - incline (2 feet long × 3 inches wide × ¾ inch thick will do fine)
    - bottom 3 feet × 3 inches × ¾ inches
    - vertical brace 1 foot × 3 inches × ¾ inch
    - small right-angle bracket to support the vertical brace

- you may want to add some kind of net to catch the marble at the end, so you don't have to chase it every time

- small flat-head wood screws—the smaller the better, but just larger than the screw holes in the track

- a meterstick

This apparatus is also commercially available, as shown in Figure 63-2.

Method

Building the track

1. Assemble the frame by attaching the pieces of wood, as shown.

2. Form a circular loop by carefully bending the track around the form. Note, the channel should be toward the inside of the circle when you do this.

3. Predrill holes for the wood screws in the wood using a drill bit a size or two smaller than your screws.

4. Secure the loop to the frame using the wood screws. It is important that the (flat) heads of the woods screws do not interfere with the

Figure 63-2 *Courtesy PASCO.*

Figure 63-3 *Courtesy PASCO.*

motion of the ball. If you find your wood screw protrudes into the path of the marble, you can work around this by enlarging the holes or by countersinking the holes in the track, so the screw head is flush with the bottom of the track.

5. Align the track as shown in Figure 63-1. The loop should be as symmetrical as possible with the overall path making a vertical loop. Also, make enough separation between the part of the track going into and out of the loop, so there is enough clearance between the marble and the track.

6. You can (optionally) attach some kind of catcher (a net or cup) to avoid chasing marbles.

Testing it

1. Take a guess as to where the marble must be placed to negotiate the loop. Here are some choices: a) equal to the radius, b) equal to the diameter, c) greater than the diameter, or d) twice the diameter. (Take into account there will be some friction.)

2. Pick your starting point and observe what happens. Find the minimum point to consistently negotiate one loop. What happens to the marble if you release it at a point that is higher or lower than this minimum point? See Figure 63-1.

Expected Results

With a low-friction sliding object car, such as a cart with wheels or a roller coaster car, the height must be at least 2.5 times the radius of the loop. Actual loops require slightly greater height to overcome friction.

For rolling objects, such as a steel ball or marble, some of the potential energy is tied up in rolling, so the height must be at least 2.7 times the radius of the loop (again, without accounting for frictional losses).

Why It Works

The potential energy you start with (by raising it to certain height on the track) is changed into kinetic energy. The higher your release point, the faster it goes. If the object is rolling rather than sliding, some of the potential energy is used to get the object rolling. If there is friction along the way, some additional potential energy is consumed.

To negotiate the loop, the centripetal force (provided by the track to maintain a circular path) must just equal the force of gravity. With less velocity, it will fall before completing the loop. With extra velocity, it will get through with some energy to spare.

Other Things to Try

Now that you have one loop down, you can try a similar track with more than one loop. You still only need one ramp to give the marble an initial velocity.

The Point

Total mechanical energy is conserved. Potential energy is converted to kinetic energy and vice versa.

Section 6

Sound and Waves

Project 64
What does sound look like? Oscilloscope wave forms.

The Idea

What does your voiceprint look like? You cannot see sound. But you can change the sound waves into electrical signals that can be displayed on a screen. Just as you found ways to visualize motion and to represent motion using various graphs, in this section you develop techniques to visually represent waves. This can enable you to study basic wave properties and to observe how waves combine to form new patterns.

You can go about this in two ways. One way is to use an oscilloscope, which is an instrument that takes an electrical signal and displays it in graphical form. Recently, a much lower cost alternative has become available that makes it possible to turn a computer into an oscilloscope.

This project focuses on how either type of oscilloscope can be used to study the wave properties of sound.

What You Need

- oscilloscopes, which range in cost from just under $600 to thousands of dollars

- sound card oscilloscope. You can turn your computer into a oscilloscope in several ways:

  - PC sound card distributed for private and noncommercial use in educational institutions at www.zeinitz.de/Christian/Scope_en.html. (Oscilloscope images shown in this and other sections are based on this sound card oscilloscope and appear courtesy of C. Zeinitz.)

  - Zelscope is available for a small charge at www.zelscope.com (this used to be called Winscope).

- tuning fork

- adapters

  - To connect microphone to computer. Microphones are either high- or low-impedance connections and the computer input is typically a mini.

  - Microphone output to oscilloscope input (typically BNC connector).

  - Depending specifically on what connections you need to make, you can most likely find connectors at Radio Shack or build the connector you need.

– Caution: Sound card oscilloscopes can handle only low-voltage inputs, such as from microphones. Attempting to use a sound card oscilloscope for larger electrical signal may damage your sound card. A reference for how to assemble a high-impedance circuit that can enable using a sound card oscilloscope for higher voltages is given in Project 115.

• wave generator

– stand-alone device designed for this purpose

– keyboard with appropriate connectors

– waveform generator available with some computer oscilloscopes

Method

Setting up the oscilloscope

1. Connect the microphone to the oscilloscope input.

2. Collect a test signal, such as your voice or a musical sound.

3. Adjust the vertical scale, so the entire wave is displayed.

4. Adjust the horizontal (time) scale, so the wave is displayed.

5. If necessary, adjust the trigger to enable the wave to be properly displayed. (Chose continuous rather than single-event settings for the trigger.)

Displaying waves

1. Generate a pitch audibly with a tuning fork, a keyboard synthesizer, or by a waveform generator. (Depending on your setup, you can use the waveform generator to produce an audible signal through a loudspeaker or send it directly into the input of the oscilloscope.)

2. Increase the pitch (frequency) and compare to the previous shape.

3. Decrease the pitch and compare to the previous measurements.

4. Increase the volume (amplitude) of the sound and observe how the wave changes.

5. Try your voice using the microphone. How does that compare to a pure tone, such as produced by the tuning fork?

6. Observe different waveform shapes, such as sinusoidal, triangular, square wave, and sawtooth. How do they sound? What musical instruments do each of the previous waveforms most closely resemble?

7. Play various musical instruments and identify fundamental waveforms that appear to be present in the instruments' waveforms.

8. Just for fun: Observe various samples of music. Can you distinguish various musical styles just by looking at the waveform?

9. Can you recognize the "voice signature" of different people as crime labs do all the time on TV?

Adding waves

1. Generate a tone or frequency. Let's say we start with 440 hertz (Hz), a concert A. Display this on Channel 1.

2. Generate a second tone or frequency. Let's say we use 100 Hz. Display this on Channel 2.

3. Many oscilloscopes let you display two signals on one display. If your oscilloscope has the capability to display two inputs on one display, show the combined signals from 1 and 2. How does the combined signal compare to the two individual signals?

4. You can also accomplish this by generating two audible tones at the same time, such as playing two notes on a keyboard synthesizer at the same time. Sounding two tuning forks at the same time will also work.

Expected Results

Increased pitch shows up on the oscilloscope as increased frequency.

Increased volume is displayed as increased amplitude.

A tuning fork or a wave generator produces a pure sine wave. Figure 64-1 shows the relatively pure sine wave pattern produced by the flute setting of an electronic synthesizer playing a 440 Hz tone.

Sawtooth and triangular waves sound more "reedy," like a clarinet or saxophone.

Other sounds are complex mixtures of simpler forms. For instance, a synthesized rock organ consists of a wider range of overtones combined with the fundamental tone. Figure 64-2 shows

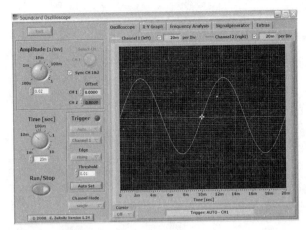

Figure 64-3 *100 Hz tone.*

several higher frequencies combined with a 440 Hz fundamental.

Adding two waveforms results in a combined sound. Figure 64-3 shows a 100 Hz tone and Figure 64-4 shows a 400 Hz tone.

Figure 64-5 shows both of these tones combined. The overall pattern shows how both of these tones add to produce a combined wave pattern.

Musical sounds are complex mixes of many individual frequencies with a large variety of overtones. Figure 64-6 is a sample from The Beatles and Figure 64-7 is an Allison Krause fiddle solo.

An oscilloscope can also show the mix of frequencies in a particular sound. For instance, a

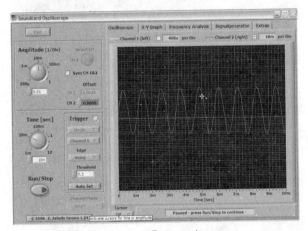

Figure 64-1 *Synthesizer flute setting.*

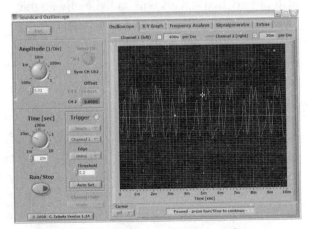

Figure 64-2 *Synthesizer rock organ setting.*

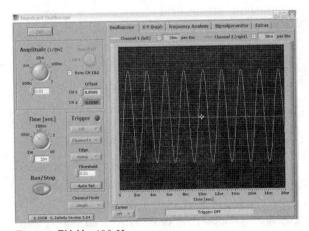

Figure 64-4 *400 Hz tone.*

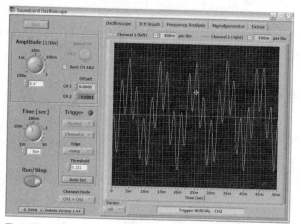

Figure 64-5 *100 Hz combined with 400 Hz tones.*

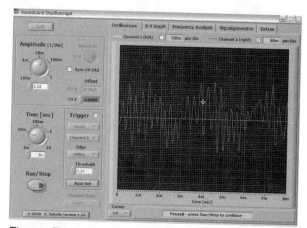

Figure 64-7 *The high-lonesome sound of a bluegrass fiddle.*

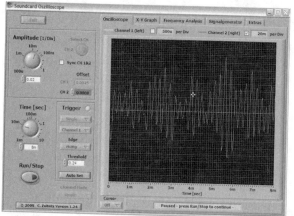

Figure 64-6 *"Eleanor Rigby" by The Beatles.*

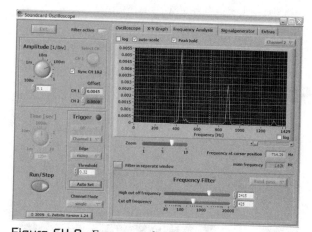

Figure 64-8 *Frequency distribution of a 440 Hz violin tone showing overtones.*

synthesizer violin sound when playing a 440 Hz tone also has some overtones at 880 Hz and 1360 Hz, as shown in Figure 64-8.

The mix of overtones contributes to establishing the musical identities of various instruments. For instance, a recorder has a very pure tone with very few overtones. Other sounds, such as a rock organ or a distorted bass, have a much more complex mix of overtones.

Why It Works

An oscilloscope processes an electrical signal and displays it in various ways. The origin of the electrical signal may be a microphone that converts a sound pattern into an electrical pattern,

which the oscilloscope can work with. The most basic form of display is a single signal versus time. The scales are adjustable to permit a wide range of signals to be displayed. Oscilloscopes also display two signals both individually on the same screen or added. A plot of one signal against the other and a distribution of frequencies are also common options.

Other Things to Try

Here is a low-tech way of picturing sound: Cover a soup can (clean, empty, and with the top removed) with Latex or other rubbery material. Put it on tight, like a drum. Attach it with a wire tie, hose clamp, or good string. Glue a small (roughly 1 centimeter on a side) piece of mirror to the top of the Latex. To use it, hold the can in one hand and shine a laser on the mirror, so the beam projects onto the ceiling (or a wall). If you don't have a laser, direct sun works as well. With the light reflecting off the mirror, create sounds that will cause the Latex to vibrate. Because of the optical geometry, the movement of the reflected laser is larger (amplified) than the smaller movement of the mirror. Because the

"drum" will be vibrating in two dimensions, it is not hard to generate the Lissajous patterns where the reflected light retraces a curved path.

The Point

Sound is a wave that, if converted into an electrical signal, can be displayed in a graphical form.

Project 65
Ripple tank.

The Idea

Water waves are probably the most tangible type of wave. For this reason, water waves can be useful in studying wave properties in general. A ripple tank provides a simple, convenient way to produce and study waves and the various types of obstacles they can encounter.

What You Need

- shallow tray or tank with a transparent bottom or commercially available ripple tank
- water
- bright light that can be held or mounted above the tank
- one or several plain sheets of white poster board to serve as a screen on which to view the images produced by the ripple tank
- various props including a straight wall, a curved wall, a thick glass plate about one-half the thickness of the water in the tank, a cup, a pencil, and a manual or mechanical source of ripples
- optional: a way to project the shadows generated, such as an overhead project or a video monitor

Method

Basic wave generation

1. Set up the tank with the light overhead. The shadow pattern should be visible on the floor.

2. Adjust the height of the light above the tank and the screen below the tank to give the best focus of the shadows from the ripples on the floor.

3. Using the tip of a ruler, tap the surface of the water to produce ripples. If you have a vibrating ripple generator, using that might give more consistent results and you won't get tired as quickly from making ripples.

4. You should see the ripples spread out in a circular pattern. The tank should be large enough, so this outward moving circular pattern is not obscured by the reflection of the ripples from the side of the tank. Sometimes, a border of foam cushioning is used to minimize side reflections.

5. Estimate the wavelength (average distance between ripples) and frequency (number of ripples per second). Estimate the velocity of the ripples. Compare this with the velocity predicted by the wave equation (which applies to all waves): velocity = wavelength × frequency (in cycles per second, which is the same as hertz). If you measure the wavelength in centimeters, the velocity will be in centimeters per second.

Reflection

1. Insert a straight barrier—a wall—in the tank.

2. Generate ripples moving toward the barrier at various angles.

3. Observe the angle the reflected waves make compared with the incoming waves.

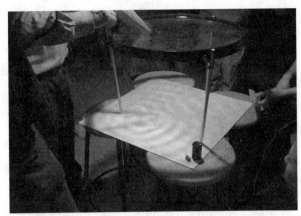

Figure 65-1 *Ripple tank showing shadows of wave patterns on white board.*

Concave and convex curved reflector

1. Insert the concave reflector. This is where the sides curve *toward* the source of the ripples. Observe how the waves are reflected. Do the waves converge or diverge?

2. Generate ripples that originate at that focal point. How do the ripples move?

3. Insert (or reshape) the reflector, so it is convex. This is where the sides curve away from the source of the ripples. Do the waves converge or diverge?

Refraction

1. Place a thick plate in the tank.

2. What happens to the speed of the waves as the waves cross over the plate? What happens to the wavelength? Does this make sense given that the frequency doesn't change?

3. Direct waves to the plate at an angle. What happens when the waves cross from the deep water to the shallower water?

Diffraction

1. Generate ripples and observe what happens when they encounter a pencil held vertically in their path.

2. What happens when a larger barrier, such as a glass or beaker, is held in the path of the ripples?

Interference

1. Generate ripples from two different locations. The ripples should be synchronized in such a way that each ripple maker goes up at the same time and down at the same time. (This means the sources of the waves are in phase.)

2. Observe what happens to the pattern as the waves from the two sources overlap and interact with each other.

Expected Results

Straight barrier: the incoming angle equals the outgoing angle.

Concave barrier: the reflected waves converge at a focal point.

Concave barrier: ripples generated at the focus regroup and emerge as a single wave.

Convex barrier: waves diverge from any location.

Plate: the waves slow as they cross over the plate; the *wavelength increases*.

Plate: the waves coming toward the plate at an angle are bent to a *less-severe* angle.

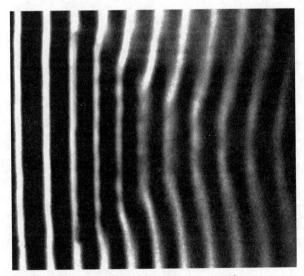

Figure 65-2 *Ripple patterns cross over the convex-shaped barrier, resulting in the convergence of the wave pattern. Courtesy PASCO.*

Figure 65-3 *This ripple tank (with vertical display) shows the diffraction pattern produced by two separate sources of wave generation. Courtesy PASCO.*

Diffraction: the wave fronts regroup around a small barrier, but not a larger one.

Interference: two ripple locations result in a fixed pattern of high and low waves.

Why It Works

Water waves exhibit basic wave properties, including:

- Reflection from straight surface: Angle of incidence equals angle of reflection (with all angles defined with respect to the perpendicular or *normal* line that can be drawn to the reflecting surface).

- Reflection from a concave surface: Waves are reflected from a curved surface with the law of reflection applying to the tangent line of the curve at that point. For approximately parabolic reflectors that include semicircular reflectors, this results in waves passing through a focal point. If the waves are generated at that focal point, they become focused and propagate in a single direction.

- Reflection from a convex surface: Waves diverge and propagate over a wider range of angles than when they started. There is no focal point when waves reflect from a convex surface.

- Refraction: Waves bend toward the perpendicular line (called the normal line) when they enter a region where the light waves move more slowly.

- Diffraction: Waves bend around a barrier in their path if the diameter of that barrier is small compared with the wavelength.

- Interference: Crests and troughs of waves combine to form an overall pattern based on constructive and destructive interference.

Other Things to Try

A large stationary body of water can serve as a large ripple tank. In this case, traveling waves can be observed without the complication of reflections from the side of the ripple tank. Pictured in Figure 65-4 is an interference pattern formed by two rocks thrown into a lake.

The Point

Waves exhibit certain characteristic behavior, including reflection, refraction, diffraction, and interference. These properties are common to all types of waves.

Figure 65-4 *Two rocks form an interference pattern as the ripples they produce spread across the surface of a lake.*

Project 66
Simple harmonic motion. The swinging pendulum.

The Idea

A pendulum undergoes a type of motion that is predictable. The consistency of pendulum motion has allowed it to be used to drive the timing mechanism of clocks. In this experiment, you investigate what causes a pendulum to swing faster or slower. At least for a pendulum on the surface of the Earth, only one variable determines the time it takes for a pendulum to swing back and forth one time.

What You Need

- several masses that can be attached to a string (such as 20 g, 50 g, 100 g, 200 g)
- several strings of varying lengths from 0.1 to 1.0 m (strong enough to support the masses)
- support for each pendulum
- stopwatch
- meterstick

Method

1. Set up a basic pendulum with a measured length and mass free to swing.

2. Pull the pendulum back to the side through a small (less than 15 degrees) angle and get the stopwatch ready.

3. Release the pendulum and start the stopwatch as the pendulum is released.

4. Count ten cycles back and forth. Cycle number one is when the pendulum returns to its original position. Be careful not to count "one" when the pendulum is released.

5. The length of the pendulum is the distance from the point where the string is supported to the center of the mass.

6. Record the time (in seconds) for the pendulum to complete ten complete cycles.

7. Divide the time for ten cycles by ten to get the time for one cycle or the period of the pendulum for the conditions you are testing.

8. You can proceed in several ways at this point, with many opportunities to develop your own plan. Here are a few suggestions:

   - What variable matters: Mass? Length? Angle? Test the selected variable while holding the others constant. For instance, test light, medium, and heavy mass, and then determine whether the period of the pendulum is dependent on mass. This can be done by measuring the period of a pendulum constructed with each of the three masses. It can also be done qualitatively by setting up three pendula and observing how fast they swing compared to each other.

   - Once you determine which variable(s) affects how fast the pendulum swings, you can set up an experiment to measure how the period changes over a range of the variables you selected. The other variables should be kept constant.

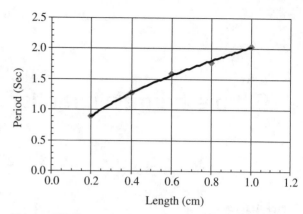

Figure 66-2 *Period versus length for a pendulum.*

A graph of period versus length is shown in Figure 66-2. The model for the graph shows the period is dependent on the square root of the length.

Why It Works

The period of a pendulum is the time it takes for the pendulum to move from one position and return to the same position. The period of a pendulum (in seconds) is given by:

$$T = 2\pi \sqrt{\frac{L}{g}}$$

where *L* is the length of the string (in meters) and *g* is the gravitational acceleration (9.8 m/s²).

This shows the dependence on the square root of the string length. Because there is no mass in the equation, the period does not depend on mass. The period also depends on the gravitational acceleration of the Earth, which under normal circumstances is not a variable.

Other Things to Try

Try this with a pendulum, consisting of a bowling ball attached to a rope. Make sure the point of attachment and the rope can securely handle the weight of the swinging pendulum.

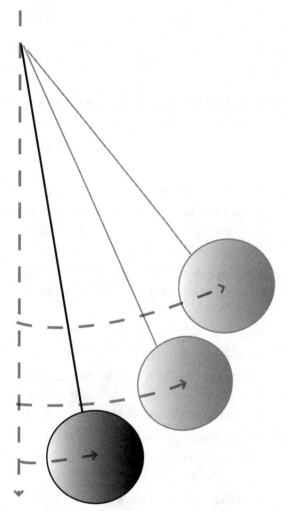

Figure 66-1 *Simple swinging pendulum.*

Expected Results

The *only* variable that affects the period of a pendulum is length. The mass does not matter at all. For angles smaller than 15 degrees, angle is insignificant. Insignificant means less than 1 percent.

The longer the string, the longer the period (*period* is the time to go back and forth one time).

The dependence of period on length is not linear.

Try this with a playground swing. Is the natural frequency of oscillation what you predict based on the previous equation? What happens if you push with a rhythm consistent with that natural frequency? What happens if you push with a rhythm very different from the natural frequency?

The Point

The period of a pendulum depends on only one variable, which is its length.

Project 67
Simple harmonic motion. The spring pendulum.

The Idea

A mass hanging from a spring is another example of a system that moves in a repeatable and consistent way. This is called simple harmonic motion. This experiment is about finding what causes a spring pendulum to vibrate faster or slower.

What You Need

- several masses that can be attached to a string (such as 20 g, 50 g, 100 g, 200 g)
- several springs of varying stiffness—it should be possible to *partially* stretch the spring by hanging each of the masses to the bottom of the spring. If the masses can't stretch the spring or if the spring is fully extended while supporting the mass, choose either other masses or other springs
- support for each pendulum
- stopwatch
- meterstick
- spring balance (for the extension)

Method

1. Set up a spring pendulum consisting of a spring with one end supporting a mass and the other attached to a support above the spring.
2. Allow the weight of the mass to stretch the spring and come to rest.

3. Pull the pendulum straight *down* through a small displacement. (Increasing the elongation of the spring by about 10 percent is a good starting point.)
4. Release the pendulum and start the stopwatch as the pendulum is released. Try to release the spring, so it goes up and down in a vertical direction. Bear in mind that, after a few cycles, a spring may have a tendency to start swinging, which complicates the type of motion we are investigating here.
5. Count ten cycles up and down. Cycle number one is when the pendulum returns to its original position. Be careful not to count "one" when the pendulum is released.
6. Record the time (in seconds) for the pendulum to complete ten complete cycles.
7. Divide the time for ten cycles by ten to get the time for one cycle. This is the period of the pendulum for the conditions you are testing.
8. As with the previous study, you can approach this investigation in several ways. You are encouraged to develop your own approach to this. Here are a few suggestions:
   - What variable matters: Mass? Spring stiffness? Amount of displacement? Test the selected variable, while holding the others constant. For instance, you can test squooshy, medium, and stiff springs, all using the same mass and displacement. (We define "squooshy" in quantitative terms in a minute.) Similarly, you can test light, medium, and heavy mass to determine whether the period of the pendulum is dependent on mass.

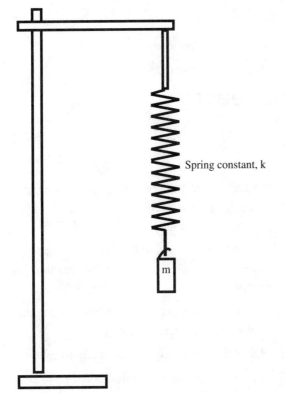

Figure 67-1 Spring pendulum.

- Once you determine which variable(s) affects how fast the pendulum swings, you can set up an experiment to measure how the period changes over a range of the variable you selected. The other variables should be kept constant.

Expected Results

The behavior of the spring pendulum is quite different than the swinging (simple) pendulum studied in the previous project.

Two variables are important for a spring pendulum: *mass* and *spring stiffness*.

The heavier the mass, the longer the period. Also, the stiffer the spring, the shorter the period. The "springiness" of a spring is called the *spring constant*, which gives a numeric measure of how stiff a spring is.

Within a fairly broad range, it should not matter whether you pull the spring through a small displacement or a larger displacement.

The dependence of period on mass and spring constant is not linear.

Why It Works

The equation for the period of a spring pendulum (in seconds) is given by:

$$T = 2\pi\sqrt{\frac{m}{k}}$$

where m is the mass (in kilograms) and k is the spring constant (in N/m). Notice the period varies as the square root of the mass and the inverse square root of the spring constant.

Other Things to Try

Predict and measure the period of the spring pendulum. You can do this by first finding the spring constant using the method of Project 30. You find the spring constant, k, by measuring the displacement, x, of a spring (in m) resulting from a given force, F (in N), according to the equation:

$$k = -F/x$$

The negative sign reflects the fact that force and displacement are always in the *opposite direction* resulting in a *positive* value for k.

Once you have determined the spring constant, predict the period of the spring pendulum using:

$$T = 2\pi\sqrt{\frac{m}{k}}$$

(The period will be given in seconds if the force is entered in *newtons* and the displacement in meters to get k. The mass must be in kg. Remember 1000g = 1kg.) Once you've called your shots, set your pendulum in motion and compare your prediction with your measured result.

The Point

The period of a spring pendulum increases as the square root of the mass. The period of a spring pendulum increases inversely with the square root of the spring constant.

Project 68
Generating sine waves.

The Idea

The simplicity of the spring pendulum provides an excellent opportunity to observe its motion in detail. The movement, like other vibrations in nature, follows a sine wave. We can also identify particular points in the spring's movement, such as where the velocity is at a maximum and a minimum during its cycle. We can also monitor how the force varies and how it relates to the acceleration. These relationships form the basis for a more complete understanding of how the various aspects of motion are interrelated.

What You Need

- spring pendulum—set up as in previous experiments

Method

1. Set the pendulum in motion and first observe when the following occurs in the cycle:
   - Zero velocity
   - Maximum velocity
   - Zero force
   - Maximum force
   - Zero acceleration
   - Maximum acceleration
2. Place a motion sensor to view the motion of the spring pendulum from underneath. If the mass presents a small target, you can tape an index card to the bottom of the mass to make it easier for the motion sensor to find. (To avoid air resistance, keep it small.)
3. Adjust the settings in the DataStudio program to give the maximum number of readings per second.
4. Open files to read simultaneously: distance, velocity, and acceleration.
5. Displace the spring and be ready to release it.
6. Press Start on the DataStudio screen to begin logging data.
7. Release the spring.
8. Record a few cycles.
9. Adjust the scales, if necessary, to best display the charts. Use the smoothing tools, if needed, to give a smoother curve if the acceleration data appears slightly choppy.

Expected Results

The equilibrium position is the point where the stationary mass hangs without moving. At the equilibrium position, the velocity is maximum, but the force (and, therefore, the acceleration) is zero.

At the maximum displacement position (the point from where the spring was released), the velocity is zero and the force (and, therefore, the acceleration) is maximum. Graphs generated by a motion sensor measuring a pendulum are shown in Figure 68-1.

Section 6: Sound and Waves

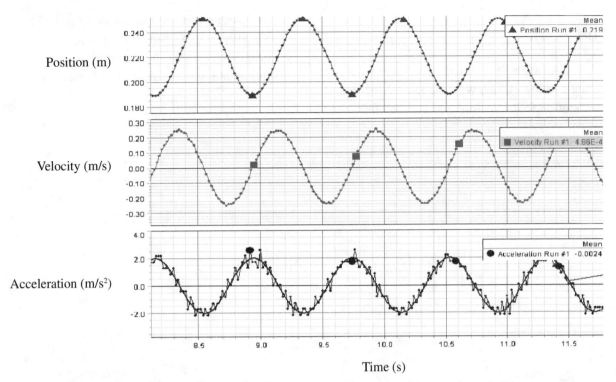

Figure 68-1 *DataStudio graphs of motion sensor data showing distance, velocity, and acceleration versus time (setup by T. Dragoiu and J. Silver).*

The distance versus time graph is a sine wave.

The velocity versus time graph is a cosine wave. The velocity is zero when the distance is at a maximum. The two waves have a similar shape, but the velocity curve is delayed by one-quarter of a period compared to the distance curve.

The acceleration curve is also a sine wave. It is at a minimum when the distance is maximum. The acceleration curve is zero when the distance curve is zero. The distance and acceleration curves have a similar shape, except the acceleration curve is delayed by one-half of a wavelength.

Why It Works

A pendulum works because the further the mass moves from equilibrium, the greater the force that returns it to equilibrium. This is the basis of any uniformly vibrating object (known as a *simple harmonic oscillator*). The response of a restoring force, such as exerted by a spring, is to produce motion that follows a sine wave. The acceleration moves in the opposite direction as the distance because the force exerted by a spring is opposite its displacement from equilibrium. This also causes the velocity and acceleration curves to be out of phase with respect to the distance.

Other Things to Try

A variation on this is to attach a force gauge to the spring to track the force along with the motion of the pendulum.

Physics alert: Those of you familiar with calculus will recognize that velocity is the first derivative of distance. Acceleration is the first derivative of velocity and the second derivative of the distance with respect to time. If the distance follows a sine curve, the velocity (the first derivative) is a cosine curve and the acceleration is a sine curve.

The Point

A *spring* is a simple harmonic oscillator whose distance follows a sine wave pattern. Velocity and acceleration follow a similar shape, but are delayed with respect to the distance curve. The velocity is at a maximum at the point of the greatest displacement. Acceleration is at a maximum at the point of greatest extension.

Project 69
Natural frequency.

The Idea

When you push someone on a swing, timing is important. If you push just as the swing has come to its highest point and is ready to begin its return, you will keep the swing going and increase its amplitude. However, if you push randomly, your efforts will be far less effective and, at times, you will tend to slow the motion of the swing. The reason for this is a swing has a natural frequency. If your pushing is at the natural frequency of the swing, the swing will resonate. This experiment explores the idea of resonance.

What You Need

- 2 ring stands
- 1⅛ inch diameter wooden dowel, 12 inches long
- 2 clamps (to hold the dowel)
- string of various lengths, from about 3 inches to 10 inches
- set of several small masses that can be attached to the string (large stainless steel nuts work well here or any attachable masses in the overall range from 10–50 g)

Method

1. Tie loops at one end of each of the strings and tie the other end to a mass. At least two of the strings should be the same length. The other should be random—some larger and some smaller than the matched pair. Avoid, however, having all the other strings half or double the size of the matched pair.

2. Slide the loops onto the dowel and spread the strings out evenly across the length of the dowel. The two matched strings should not be next to each other.

3. Attach the dowel to the two upright posts of the ring stands using the clamps. Leave enough space, so all the masses are free to swing without hitting the table, which the ring stands are placed on. The dowel should be slightly flexible, but constrained by the ring stands, so it will not sway or swivel when the masses are swinging. See Figure 69-1.

4. Steady all the masses hanging on the strings.

5. Take only one of the masses on the matched strings and displace it, so it is swinging perpendicular to the direction of the dowel.

Expected Results

What we want to see here is for the stationary string, which is the same length as the one that was set in motion, to also start moving back and forth. The other masses might jostle around a bit, but they should not be set into a significant swinging motion.

Why It Works

The resonant frequency of a pendulum is determined exclusively by the length of the string

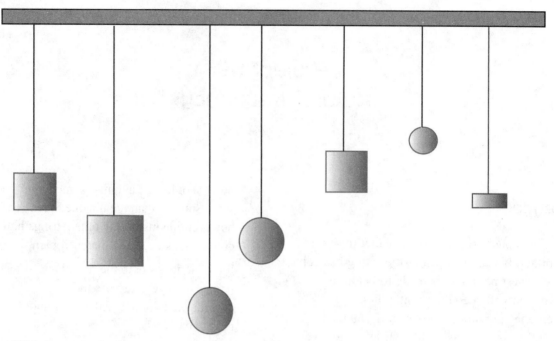

Figure 69-1 *Resonant frequency depends only on string length.*

that supports it. The stationary pendulum is stimulated by the swinging pendulum that has the same length and, therefore, the same natural frequency.

Other Things to Try

Other questions that can be addressed are:

1. Would a pendulum with the same string length, but different mass, have the same resonant frequency as the swinging pendulum?

2. What if we throw in a few harmonics? What is the response of a pendulum that is one-half the length of the swinging pendulum? What is the response of a pendulum that is double the length of the swinging pendulum?

The Point

A simple harmonic oscillator, such as a swinging (simple) pendulum, has a natural frequency. If stimulated at that natural frequency, the amplitude of that pendulum will be greatest.

Project 70
Bunsen burner pipe organ. Resonant frequency.

The Idea

If you blow across the top of a soda bottle, you produce a sound. The more soda you drink, the deeper the pitch of the sound. This is because the resonant frequency of the bottle increases as the height of the air above the liquid increases. This project does the same thing, except on a much bigger scale. You use longer tubes that produce deeper sounds. You may just want to give the person who is responsible for the building you are in (such as the building principal) advance warning that the sound they will be hearing is not an earthquake, not a water buffalo in labor, and not the propulsion system for a space alien spacecraft.

What You Need

- large Bunsen burner
- cylindrical tubes roughly 0.8–2 meters in length—good candidates for this are cardboard tubes used for carpet rolls or hollow metal sections of old driveway basketball backboard supports
- fire extinguisher and/or bucket of water (you should *not* need this, but just in case)

Method

1. Place the Bunsen burner on the floor.
2. Route a hose long enough not to get tangled and connect it to a natural gas outlet (which, at this point, is not turned on).

3. Position a cylindrical tube on the floor, so it can easily be placed over the burner.
4. Light the burner and adjust it, so the flow is maximum, with the greatest flame.
5. Make sure nothing flammable is near the burner, including possible objects on the ceiling and loose papers that might inadvertently be drawn into the flame by convections.
6. Lift the cylinder with both hands and position it above the burner. Hold it a few inches above the top of the burner, but not so low that it constricts the air going into the burner. See Figure 70-1.
7. Be somewhat careful to avoid charring the edge of the tube. Do not hold the edge of the tube directly over the flame. If held correctly, the tube will not burn. Also exercise similar care to avoid excessively heating a metal tube, which could possibly result in a burning hazard.
8. It may take as much as a minute or so until a sufficient stream of exhaust from the Bunsen burner flows through the tube to produce an acoustic resonance.
9. Measure (or, if you prefer, observe) the pitch (or frequency) of the sound produced using the pitch gauge or oscilloscope. You can also compare the pitch with a known frequency, such as produced by hitting a tuning fork or sounding a musical instrument.
10. Remove the cylinder and don't forget to turn off the burner when you are finished.

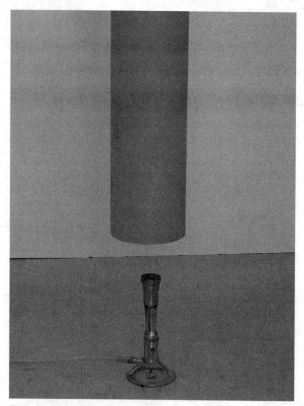

Figure 70-1 *Resonating tube.*

Expected Results

With a sufficient flow of heated air, the tube should resonate. The taller tubes produce a much lower pitch than the shorter tubes. The diameter of the tube does not affect the pitch, but it may affect the volume by impacting how much air can flow.

For a given length, the frequency of the tube is given by:

Column length (m)	Resonant frequency (Hz)
0.2	440
0.5	172
1.0	86
1.5	57
2.0	43

The shorter columns produce tones consistent with the sounds of common concert instruments. The longer columns produce very deep, resonant pitches.

Why It Works

The flowing air, just like a wind instrument, creates a resonant standing wave in the tube. The longer the tube, the longer the wavelength, but the shorter the frequency.

The frequency for an open tube is given by $f = v/2L$. For a given tube length, this frequency is twice as high as it would be for a closed tube. Refining this to take into account the slight offset of the node from the ends of the tube, the equation is:

$$f = v/2(L + 0.8d)$$

where d is the diameter of the tube (in m), L is the length of the tube (in m), and v is the speed of sound (in m/s).

Other Things to Try

If you have more than one tube whose diameters are similar, you can nest several tubes together, one inside the other, as shown in Figure 70-2. This can let you continuously vary the pitch of the tube, like a trombone. You may want to practice before auditioning for the spring musical.

The Point

The resonant frequency in an open pipe is lower for a greater length.

Figure 70-2 *Changing the note like a trombone.*

Section 6: Sound and waves

The Idea

On one side of the room, you have a bar magnet suspended from a spring. The magnet is surrounded by an electrical coil, which is attached to another coil on the other side of the room. An identical magnet is placed inside the second coil. As the first coil starts moving, an electrical current is produced, which also causes the second magnet to start moving. This is impressive to see and demonstrates several principles of physics, including resonance, magnetic induction, and magnetic force.

What You Need

- 2 bar magnets
- 2 equal springs
- 2 wire coils with an interior opening just large enough for the magnets (these can be made or are available from scientific equipment supply companies)
- 2 ring stands with clamps to support a horizontal connecting bar; 2 pendulum clamps would be perfect
- string
- table
- connecting wire
- optional: 4 LEDs

Method

Overall, you are going to set up two identical parts of the apparatus, shown in Figure 71-1, connected together electrically. To do this, follow these steps:

1. Suspend each of the two springs to the supports.
2. Attach the two bar magnets to the bottom of the springs using string or wire.
3. Position the bar magnets, so when the spring is displaced downward, the magnet extends into the coil, but it does not touch the table the coils are sitting on.
4. Start with both springs hanging.
5. Displace one spring to set it oscillating, but leave the other spring hanging undisturbed. Observe the result.

Expected Results

Initially, the first magnet, after being set in motion, goes up and down by itself. The motion of the first magnet generates an electrical current that causes the second magnet/spring combination to start to oscillate.

Why It Works

The first magnet moving through the coil generates a current. This current is transmitted to the second coil. The current flowing in the second

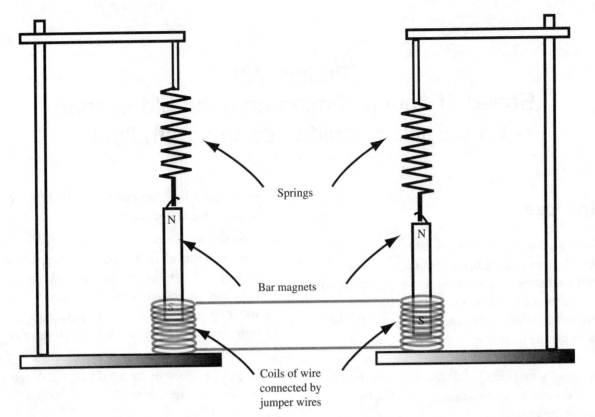

Figure 71-1 *The movement of one spring causes the other one to resonate.*

coil exerts a force on the second magnet, which sets it in motion. Because both springs are a matched set with nearly identical spring constants, the frequency of the electrical signal driving the second spring is at its resonant frequency. A small driving force at the resonant frequency has a much greater impact than a force at any other frequency.

Other Things to Try

If you want to push your luck, you can put an LED in the electrical circuit. LEDs conduct current in only one direction. A pair of LEDs, each oriented in the opposite direction and connected in parallel, would be needed to prevent blocking the current flow. You can also put a galvanometer or a current sensor in series with one of the wires and measure the current flow directly.

Another simple way to show oscillation between magnets is to suspend two magnets horizontally from springs. Start with the north pole of one magnet facing the south pole of the other magnet. Then, turn each of the magnets from that equilibrium line. They can be turned through an angle in the same or opposing directions. The magnets will move to bring themselves back to that equilibrium position. But they will overshoot and keep going until their energy is lost to friction. Until then, they form a simple harmonic oscillator. Try this with different starting angles.

The Point

A magnet moving in a coil of wire generates an electrical current.

An electrical current moving in a wire exerts a force on a magnet.

A simple harmonic oscillator (in this case, the spring) resonates if driven at its resonant frequency.

Project 72
Speed of sound. Timing an echo old school.
Why Galileo couldn't do this with light.

The Idea

The speed of sound, like any other sound, can be found by measuring the time it takes to go a certain distance. This simple and straightforward measurement can give a reasonable ballpark estimate, but not highly accurate results. We will, however, be limited by the large distances we need to work with and the small times we need to accurately measure. We measure the speed of sound using a more accurate method in the following projects.

What You Need

- long tape measure (or some other way to *estimate* a long distance, such as counting cinder blocks of a known length on a building or clocking the distance of several blocks using the odometer of a car)

- means of generating a loud sound (such as an air horn, garbage-can lid, or baseball bat, or someone with a loud voice)

- stopwatch

- partner (which may not be needed if you can set up an echo)

Method

1. Measure or estimate a course of known or estimated distance without visual obstruction. A football field or possibly multiple lengths can work. You can also use a building or natural geographic feature, such as a cliff to reflect an echo. This effectively doubles the distance the sound travels.

2. Generate the sound and note the difference in time between when the sound was generated and when it is heard at a distant location. (This can be accomplished by observing when the garbage-can lid was struck or observing at a distance when the air horn is sounded.)

3. To get the speed of sound, divide the distance by the time. See Figure 72-1.

Expected Results

The speed of sound is about 343 meters per second or about 1096 feet per second at 20°C. It is unlikely this technique will give an accurate value for the speed of sound, but it should provide a ballpark estimate.

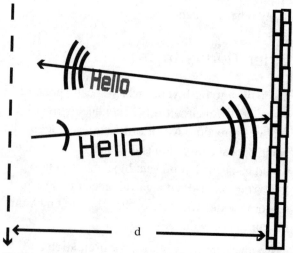

Figure 72-1 *Measuring the velocity of sound directly.*

Why It Works

Velocity is distance divided by time. Because the speed of light is so much greater than the speed of sound, the time it takes light to travel the distance can be considered essentially zero and is insignificant compared to the speed of sound.

Note that Galileo tried to measure the speed of light using a similar method. Light moves so quickly, however, it requires extremely large distances to measure the time it takes to travel using a stopwatch. Rather than saying that Galileo failed in his attempt, we like to say he succeeded in proving that light was much faster than he could measure.

Other Things to Try

The distance to a lightning strike can be determined by counting the number of seconds between seeing the lighting and hearing the thunder. Using a known value for the speed of sound multiplied by time can provide an estimate of the distance to the lightning strike. Similarly, the speed of sound can be determined if the distance to the lightning strike is known (or can be measured, such as by driving to where the lightning was observed to hit) and divided by the time between seeing the lightning and hearing the thunder.

The Point

Speed is distance divided by time. However, the accuracy of any experiment is limited by the resolution of the least certain measurement. Even if distance can be measured accurately, the time measurements are limited by the reaction time of the observer.

Project 73
Speed of sound. Resonance in a cylinder.

The Idea

In this experiment, we measure the speed of sound based on the resonance that a tuning fork produces over a column of water in a cylinder. Unlike the very direct approach of the previous section, we take advantage of the wave properties of sound to make a much more accurate measurement.

What You Need

- tuning fork of known frequency
- rubber mallet
- closed watertight cylinder (about 15–20 cm tall)—a 1–2 L graduated cylinder will work
- ruler
- 200 mL beaker (or other container with a spout to pour water into the graduated cylinder)
- water
- quiet room

Method

1. Strike the tuning fork with the rubber mallet. Hitting the tuning fork on a hard surface may result in altering its frequency.

2. Hold the ringing tuning fork over the top of the graduated cylinder, as indicated in Figure 73-1.

3. Place your ear near the top of the graduated cylinder and listen to the sound of the tuning fork.

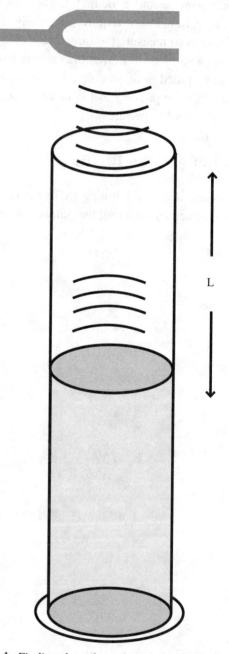

Figure 73-1 *Finding the column length that results in resonance at a particular frequency.*

4. Slowly add water to the cylinder and continue to listen. This can be a several-person operation. Be careful not to cause the water to splash, which can distract the listener from hearing the sound of the tuning fork.

5. At a certain height, as the water level is raised, the sound of the tuning fork becomes markedly louder. You may need to listen carefully to hear it.

6. Once you think you found the resonance, pour out some of the water and confirm that the sound level for the tuning fork becomes lower, and then gets louder again as the water level is brought back up. It should also get lower as the water level is raised above the resonance level. (If you miss the first resonance, continue adding water andyou will hear the second resonance. Using the wavelength for the second resonance will result in a speed of sound that is half the correct value.)

7. Determine the frequency of the tuning fork, either by noticing the marking on the tuning fork or by measuring it. You can use an instrument tuner to measure or verify the frequency of the tuning fork.

8. Calculate the speed of sound using this equation:

$$v = 4\,Lf$$

where L is the length of the air column above the water and f is the frequency marked on the tuning fork.

9. A more exact expression which accounts for the node of the sound wave not being exactly at the opening of the tube of diameter, d, is:

$$v = 4f(L + 0.4d)$$

Expected Results

The accepted value for the speed of sound at 20°C is 343 m/s. The warmer the air, the faster the speed of sound, according to the equation $v = 331$ m/s $+ 0.6\,T$ where T is the temperature in degrees centigrade.

To get an idea of the appropriate height needed in the resonant air column for various common tuning forks, you can check the following table. These values do not include the correction factor used in the experiment to account for the diameter of the cylinder. It is possible that someone doing this experiment may not notice the resonance of the fundamental frequency and will continue filling the graduated cylinder until arriving at the resonance for the second harmonic. Table 73-1 serves as a guide to find the column height that produces a resonant frequency.

Why It Works

The resonance occurs when the length of the column produces a natural resonance that is the same as the tuning fork.

Table 73-1

Approximate note	Frequency (Hz)	Wavelength (m)	Column height for fundamental (m)	Column height for 2nd harmonic (m)
C	256	1.34	0.33	0.17
D	294	1.17	0.29	0.15
E	330	1.04	0.26	0.13
G	392	0.88	0.22	0.11
A	440	0.78	0.19	0.10

One way to think about this is that, at the speed of sound, the sound goes to the bottom of the column and back in the same time as the tuning fork goes through one vibration.

Another way to say this is the frequency of the standing wave that resonates equals the frequency of the tuning fork. The wave equation states that the velocity of a wave equals its wavelength times its frequency. The equation is

$$v = \lambda f$$

where v is the velocity (in m/s), λ is the wavelength (in m), and f is the frequency (in cycles per second or Hz).

Other Things to Try

Compare this method to the other methods of measuring the speed of sound found in this book.

The Point

The speed of sound can be determined by measuring the length of an air column at which a tuning fork establishes a resonance. One-quarter wavelength fits in the air column. Measuring the air column can then determine the wavelength of the sound. This combined with the frequency determines the speed of sound.

Project 74
Racing against sound. Doppler effect.

The Idea

Weather forecasters use the Doppler effect to detect wind shear. Astronomers use it to determine that distant galaxies are moving away from each other in an apparent expansion of the universe. This experiment demonstrates how the frequency of a sound moving toward or away from a listener is affected by the Doppler effect.

What You Need

- 1 meter length of string
- 1 electric buzzer or other source of a sustained note (the buzzer must have a point of attachment for a piece of string and must not require someone to continuously activate a switch to make it sound)
- 2 people
- optional: microphone, oscilloscope, or sound card oscilloscope

Method

1. Securely attach the buzzer to the string.
2. Turn on the buzzer.
3. One person spins the buzzer in a circle, moving toward and away from the second person, as shown in Figure 74-1.
4. The observer should listen to the sound the buzzer makes as it comes toward and away from where they are located.

Expected Results

The *pitch* of the sound increases and decreases at a rate established by the period of the rotating buzzer. The volume of the buzzer sound may also increase and decrease, but that is *not* the Doppler effect. The faster the buzzer spins, the greater the difference in pitch.

The pitch is higher as the sound moves toward you and lower as it moves away from you.

Why It Works

When sound is coming toward you, the peaks and troughs of the waves are closer together, as indicated in Figure 74-2. This results in a higher frequency of the sound wave. From the perspective of the listener, the sound waves seem to come more frequently and are perceived to have a higher pitch.

Other Things to Try

Attach the microphone to either a physical oscilloscope or a sound card oscilloscope.

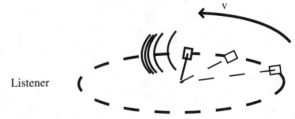

Figure 74-1 *Sound varies in pitch as it moves with respect to the listener.*

191

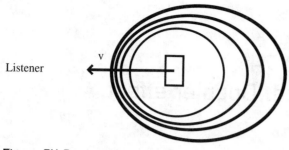

Listener

v

Figure 74-2 *As the buzzer moves toward the listener, the perceived pitch of the sound is higher. As it goes around the circle and moves away from the listener, the pitch becomes lower.*

Display the sound and compare the frequency produced by the buzzer coming and going.

Attach a buzzer to any object that can move in a more or less horizontal direction (such as an air track glider, a frictionless cart, or even an old skateboard). As the object moves, listen to the sound. If you can, also try displaying the waveform on an oscilloscope.

The Point

The Doppler effect occurs when the source of the sound is either moving toward you or away from you. When the source of sound is moving toward you, the pitch or frequency of the sound is higher; when the source of sound is moving away from you, the pitch is lower.

Project 75
Adding sounds. Beat frequency.

The Idea

When waves meet up at the same point in space, the waves add together to form a new wave. This combining of waves is called *superposition*. The waves add together in a process called *interference*. If the crests form at the same place, we have constructive interference and the combined wave is smaller. If a crest meets a trough, we have destructive interference and a smaller wave.

Sometimes when waves combine, the pattern they produce is itself a wave. We can hear the beat frequency of a sound wave most easily when two sound waves are separated by a small frequency difference.

What You Need

- source of two tones that differ by a few Hz. Some options for this include:
    - adjustable tuning fork pair with resonant cavities
    - 2 matched tuning forks, one of which can be detuned by applying a small mass to the tines of one of the tuning forks
    - (Polyphonic) keyboard synthesizer
- waveform generator with two channels or two waveform generators
- optional: a good pair of ears
- optional: an oscilloscope—either a physical instrument or a sound card oscilloscope

Method

1. Play two tones at the same time that are different by only a few Hz. For instance, you can use 440 and 445 Hz. Or, you can play two notes on a keyboard separated by a step or two.

2. Listen carefully. See if you can distinguish the first tone and the second tone individually. Then, listen for a fading in and out of the overall sound. That throbbing of the basic tone is called the *beat frequency*. The pulsation itself has a frequency equal to the difference between each of the two individual frequencies.

3. Using the technique developed in Project 64, display the combined waves for each of the tones on the oscilloscope.

4. Measure the frequency of the overall pulsating wave pattern that envelops both waves. Compare that to the difference in frequency for each of the two individual waves.

Expected Results

You should hear a pulsating throbbing tone that causes the combined tones to periodically grow louder and softer.

As an example, by combining a 440 Hz wave with a 445 Hz wave, you get a combined tone that gets louder and softer every five seconds, as shown in Figure 75-1. The beat frequency is the difference between the two original waves.

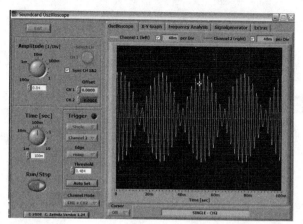

Figure 75-1 *Addition of two frequencies produces a beat frequency.*

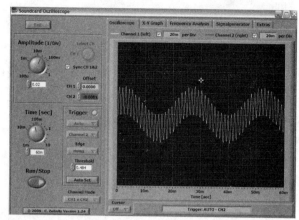

Figure 75-2 *Multiplying the amplitudes of two sound waves shows the beat frequency more prominently.*

Why It Works

When two waves are produced at the same location, the beat frequency equals the difference between the frequencies of the two waves.

Other Things to Try

We can also look at the product of the two waves that exaggerates the overall pattern of the beat frequency, as shown in Figure 75-2. Many oscilloscopes display the product of the two input waveforms.

The Point

The beat frequency is the difference between the frequencies of the two individual waves.

Project 76
Pendulum waves.

The Idea

This demonstration uses an apparatus built from several different masses hanging from strings. Each pendulum is slightly shorter than its neighbor. Because the period of a pendulum is longer for longer string, each pendulum will go back and forth in slightly less time than its neighbor. This difference results in an overall changing pattern of standing waves and traveling waves.

What You Need

- 8–12 small uniform masses (nuts, hooked masses)

- string or fishing line

- frame, as shown in Figures 76-1 and 76-2, which allows the string for each successive mass pendulum to become progressively larger

- This apparatus is also commercially available from Edmunds Scientific (item number 3123752).

The following equation (from *Pendulum Waves: The Physics of a Set of Tuned Pendulums*, Brad De Gregorio, found at member.cox.net/brad.degregorio/PendulumWave.pdf) gives the optimal lengths for each pendulum string:

Length of nth pendulum string:

$$L(n) = 9.8\left(\frac{T_{max}}{2\pi(k + n + 1)}\right)^2$$

where, T_{max} is the period of the longest pendulum, k is the number of cycles the apparatus goes through before repeating its pattern, and n is the number of the pendulum ($n = 1$ is the first pendulum, $n = 2$ is the second, and so on).

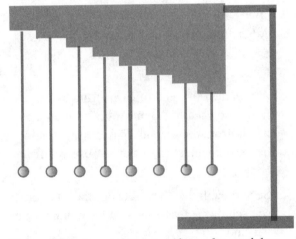

Figure 76-1 *Inverted staircase frame for pendulum wave.*

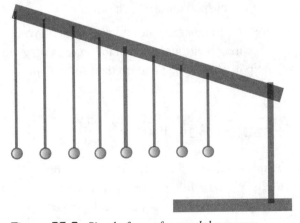

Figure 76-2 *Simple frame for pendulum wave.*

Table 76-1

Pendulum	1	2	3	4	5	6	7	8	9
Length, L(n) in cm	35.7	33.0	30.6	28.5	26.6	24.8	23.2	21.8	20.5

Table 76-1 gives the length of a set of strings for a pendulum wave to repeat every 30 seconds with 25 oscillations for the longest string during that time.

Notice that the optimal string length is not linear. Some pendulum wave frames are actually curved to accommodate the shape given by the previous equation.

Method

1. Attach the strings to the masses.
2. Adjust the string lengths between the masses and the frame, so all the masses are the same length.
3. Attach the other end of each string to the frame, so each of the masses is at the same height from the ground. The masses should be low enough so they can be observed from above.
4. Secure the frame to a table or other supports, so each of the masses is free-swinging above the floor.
5. Hold all the masses and push off to one side using a meterstick or flat board. The masses are not pushed toward each other.
6. Release the masses and observe from above. Placing a board underneath the masses may help in viewing the pattern they form as they move.

Expected Results

The masses define a continuously changing pattern. With the first swing, the masses all swing pretty much together, resulting in the patterns shown in Figures 76-3, 76-4, and 76-5.

Figure 76-3 *Linear alignment.*

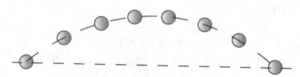

Figure 76-4 *Quarter wave.*

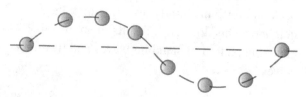

Figure 76-5 *Full wave.*

Why It Works

The period of a pendulum, or the time, T (in seconds), it takes to swing back and forth one time increases with the length of the pendulum, according to the formula:

$$T = 2\pi(L/g)^{1/2}$$

where L is the string length (in meters) and g is the gravitational acceleration constant (in m/s^2).

Because each successive mass has a slightly longer string, its period is longer than the mass before it. The more cycles the masses go through, the more difficult it is for the masses on the longer strings to keep up. The delay that occurs in the slower masses begins to develop into the patterns depicted in the previous figures.

Other Things to Try

A follow-up to this exercise is to predict the period during which a sequence of patterns repeats. This can be verified by measuring how often a particular pattern takes to recur compared with the step size for each successive pendulum.

The Point

This project illustrates the variability of the period of a pendulum with string length. It also shows how changes in the frequency of a wave can have an effect on whether it is in phase or out of phase with other masses in the system.

Project 77
Using waves to measure the speed of sound.

The Idea

In this experiment you will determine the speed of sound by measuring how long it takes sound to get from one microphone to another separated by a known distance. This is almost the same thing you did in Project 72. The only difference is that here we use an oscilloscope to measure the difference in time instead of a stopwatch.

You can take advantage of the wave properties of sound to find the distance between the positions where the sound is loudest. This occurs where the sound constructively interferes. This lets you find the wavelength of the sound. Knowing the wavelength and frequency of sound lets you determine its velocity.

This experiment provides an opportunity to explore basic properties of waves in general. The overall techniques used here can, with significant refinements, also be used to measure the speed of light.

What You Need

- 2 speakers
- 2 approximately 6-foot lengths of hookup wire
- tone generator or a single tone *wav* file played through a computer or digital audio player
- 2 microphones connected to an oscilloscope (or a sensitive sound meter)
- tape measure, meterstick
- quiet room

Method

Two speakers/one microphone

1. Connect the tone generator to the two speakers using the hookup wire. Connect the positive terminal of the tone generator to the positive terminal of *each* of the speakers. The negative terminal of the tone generator is connected to the negative terminals of *each* of the speakers.

2. Position both speakers side-by-side directed toward the microphone. At this point and throughout this measurement, each speaker should have an unobstructed line-of-sight view to the microphone, as shown in Figure 77-1.

3. Turn on the tone generator. Verify that both speakers are functional and at roughly the same volume. You should hear a steady, continuous tone. Any midrange range frequency should work, such as 440 Hz, although this method works well for all audible frequencies.

4. Connect the microphone to your oscilloscope. (Alternatively, you can use a sound meter or just listen carefully to determine the positions of maximum and minimum sound intensity.)

5. Display the waveform picked up by the microphones on the oscilloscope. Adjust the amplitude, time scale, and, if needed, the trigger setting.

6. Slowly move one of the speakers (either forward or back) along the line between it and the microphone. Each speaker should, at all times, continue to face the microphone.

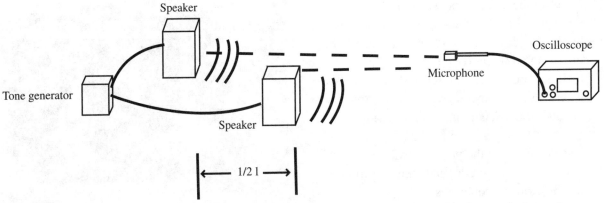

Figure 77-1 *Finding the distance between positions of maximum intensity to determine the speed of sound.*

7. Monitor the amplitude of the signal displayed on the oscilloscope (or the intensity on the sound sensor; you can also hear the relative intensity of the sound with reasonable accuracy). Be careful to avoid any objects that could block or reflect the sound waves striking the microphone.

8. Note the frequency of the sound waves (from the setting on the tone generator or the *wav* file you used). However, if you don't know the frequency, or just want to confirm it, determine how many seconds it takes on the time scale for one full oscillation to occur. The time it takes for one wavelength to occur is called the *period* of the sound wave. The *reciprocal* of the period is the *frequency, f* (in Hz or cycles per second).

9. As you adjust the distance between the speakers, you should see the amplitude of the combined sound waves decrease, reach a minimum, and then return back to its maximum level as the speakers are moved.

10. The *distance between the speakers* when the sound is at a *maximum* is a *full wavelength*. This is the result of constructive reinforcement of the signals. The distance between the speakers when the sound is at a *minimum* is a half wavelength, resulting in destructive interference. (The components for this experiment are shown in Figure 77-2.)

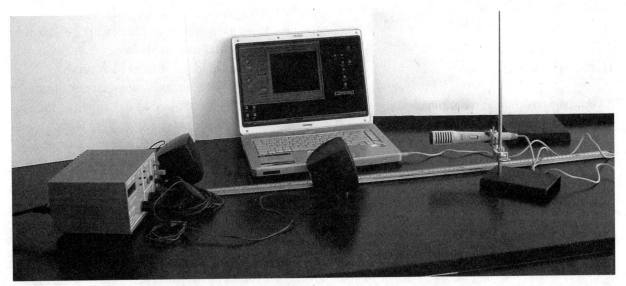

Figure 77-2 *Components to measure the speed of sound.*

11. Measure the distance between the two speakers when the sound is at maximum level. This distance is one full wavelength, λ, of the sound wave. If you measure this in meters, your calculation for the speed of sound will be in meters per second. (You can get additional data points by measuring different locations, finding the one-half wavelength from the positions when the sound is at a minimum, and then repeating this at various frequencies.)

12. Once we have the wavelength, λ, and the frequency, *f*, you can multiply them together to get the velocity using the wave equation:

$$v = \lambda f$$

Expected Results

As before, the speed of sound at 20 degrees centigrade is 343 meters per second.

The speed of sound (in meters per second) as a function of temperature (in degrees centigrade) is $v = 331 + 0.6T$.

Using a 440 Hz tone, the distance separating the microphones to get a 343 meter per second value for the speed of sound is 0.78 meters (78 centimeters).

A 1000 Hz tone would require a 0.343 meter separation to result in the expected value for the speed of sound.

Why It Works

Two waves traveling in the same direction add together to form a new wave. If the crests of the two waves rise at the same time and place, the waves are said to be in phase and reinforce each other to produce a louder sound. This is called constructive interference. This occurs when the two sources of the sound are separated by exactly one full wavelength. (One wave gets a one-wavelength head start, but both are in phase at the detector.) If we know the wavelength and the

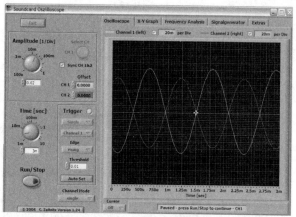

Figure 77-3 *Sound waves (amplitude versus time) shifted by one-half a wavelength as a result of traveling through the distance between the two speakers.*

frequency of the sound, we can easily determine its velocity according to the relationship $v = \lambda f$.

Destructive interference occurs when one wave crests while the trough of a second wave is passing. This happens when the source of the two waves is separated by half a wavelength.

Other Things to Try

Two microphones/one speaker

If you can set up two microphones to your oscilloscope, there is another way to do this that shows the process of interference more clearly. In this case, you follow basically the same procedure as the previous one, except you have one speaker and two microphones. You move the microphones until you observe destructive interference. This occurs when the crest of one wave is at the same place as the trough of the other wave, as shown in Figure 77-3. This method does not work using a sound intensity meter or by listening carefully, as did the previous method.

Following this method, you can use the capability that many oscilloscopes have to detect the point at which the waves are separated by one-half a wavelength. This involves plotting one signal versus the other on the display. When this

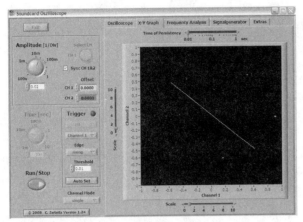

Figure 77-4 *x–y plot of two sound waves showing the separation of one-half wavelength.*

x versus *y* plot is a straight line with a slope equal to a negative one, as shown in Figure 77-4, your signals are 180 degrees out of phase and separated by one-half wavelength.

Interference along a line/ double slit analogy

Another configuration that can be used to find positions of a constructive and destructive interference is shown in Figure 77-5. This method is analogous to the double slit technique used by Thomas Young with light and is explored in Project 83. The wavelength is given by:

$$\lambda = d\sin\theta$$

where *d* is the separation between the speakers and θ is the angle between the midpoint between the speakers and the point where constructive interference is identified.

Speed of light

Using a similar principle, the speed of light can also be measured in the lab. An apparatus is commercially available that determines the wavelength of a known frequency of light by measuring the distance between positions of constructive interference. The measurement is much trickier than the one in this experiment because the speed of light is so much greater than the speed of sound. The same basic approach, however, can be applied to either sound or light.

The Point

The speed of sound can be determined if the wavelength and frequency of the wave are known. The wavelength for a given frequency can be determined by finding the distance at which constructive interference occurs.

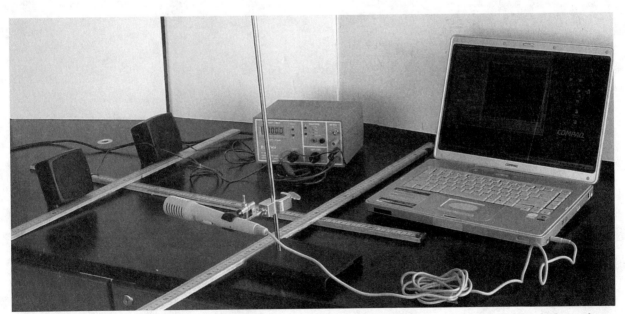

Figure 77-5 *Finding the locations of maximum and minimum sound intensity to determine the speed of sound.*

Light

The Idea

This is a perfect way to see for yourself how light moves when it encounters mirrors and lenses.

What You Need

- laser pointer
- set of lenses, including convex, concave, rectangular, and semicircular lens; rectangular prism and 60 and right-angle prisms (90°, 45°, 45°) and (90°, 30°, and 60°)
- 2 small flat mirrors
- sheet of paper (plain or gridded)
- ruler
- protractor
- dark room

Method

1. Caution—You should use a low-power laser pointer and be careful *not* to shine the laser where it could hit anyone's eyes. Remember, you are working with optical devices that change the path of the light, so be careful to avoid stray light rays that could affect anyone's eyes.

2. Place your object lens (or mirror) on a flat table.

3. Place a sheet of paper underneath the lens.

4. Trace the outline of the lens on your paper. Leave enough room to draw incoming and outgoing lines.

5. Darken the room.

6. Shine the laser at a slight angle, so its straight line path can be seen on the paper.

7. For each of the lenses, put three or more dots along the path to define the incident path.

8. Observe its path through the lens (or reflected from the mirrors). You may need to slightly adjust the angle (to the plane of the table) to make the transmitted ray visible. Depending on your lenses, you may not be able to see the laser light going through the lens. Also, be careful not to mistake light that may sneak underneath the lens as a ray that follows the intended optical path through the lens. Also (again depending on your lenses), you may need a slightly different angle to make the

incident laser line visible as you would need for the refracted line. If this is the case, make sure you come into the lens along the same incident line that you drew.

9. Make three or more dots to define the refracted (or reflected) paths.

10. Make a dot where the light enters the lens and where it leaves the lens.

11. Connect the dots with straight lines showing the incident (incoming) line, the straight line through the lens (which is the refracted line), and the transmitted or reflected lines.

12. Explore as many of the following optical objects as you have available. The following lists several specific things to focus on.

(Note: All this can be done, if you prefer, on a magnetic chalkboard using lenses with magnetic backs. You can either glue strong magnets to your lens or simply hold the lens to the chalkboard. Make sure that the magnetic chalkboard is strong enough to hold the lens securely and that the magnet does not block the path of the light. Laser levels may be useful because they have built-in angles to make the line visible along a surface. However, they may be a little trickier to focus all the way through the lens.)

Single mirror

Draw a perpendicular line to the surface of the mirror. Shine the laser at the point where the perpendicular line meets the mirror. Place dots along the incident line and the reflected line, and then connect the dots. Compare the incident angle with the reflected angle. Try this for several sets of angles (Figure 78-1).

Mirrors at a right angle

Shine the laser on one of the mirrors and trace its path (by placing dots along the path and connecting them). The path should hit the second mirror, and then reflect off the second mirror. Try this for several angles of incidence on the first

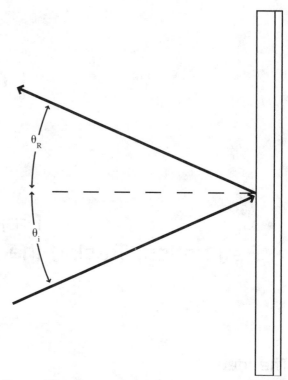

Figure 78-1 *Reflection from a plane mirror.*

mirror. (If you like geometry, set the mirrors at an acute angle. Then, predict and test the angle of the outgoing ray for a given angle of incidence.)

Convex lens

A *convex lens* is the one that is thicker at the middle than at the ends. Draw a centerline

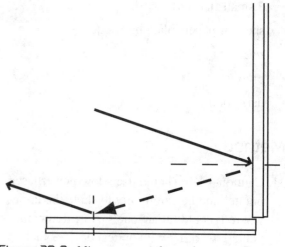

Figure 78-2 *Mirrors at a right angle.*

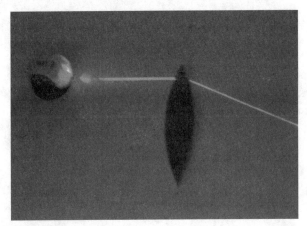

Figure 78-3 *Convex lens.*

Figure 78-4 *Concave lens.*

perpendicular to the axis of the lens. Trace the following paths: a) a straight line along the center line through the center of the lens, b) a line above the centerline running parallel to the centerline, c) a line below the centerline running parallel to it. Trace all the lines. Notice where the three lines cross. Measure that distance and put a dot on the centerline on the incident side of the lens that same distance from the lens. Direct the laser at any angle through that dot and trace its path through the lens. Try this for several angles.

Concave lens

A *concave lens* is the one that is thinner at the middle than at the ends. Draw a centerline perpendicular to the axis of the lens. Trace the following paths: a) straight line along the center line through the center of the lens, b) a line above the centerline running parallel to the centerline, c) a line running below the centerline running and running parallel to it. Trace all the lines. How do these results compare with those from the convex lens?

Semicircular lens

The *semicircular lens* has one circular side and one flat side. Place the circular side toward you. Trace the lens and draw a centerline on the flat side. Shine the laser at a 30-degree angle to that centerline, but hit the point where the centerline

meets the flat side of the lens. This particular arrangement avoids refraction in the glass because the light comes in perpendicular to the tangent at the circular edge. In this case, the only refraction that occurs is at the glass-to-air interface. Observe what happens for different angles. Take it to the extremes of high and low angles of incidence.

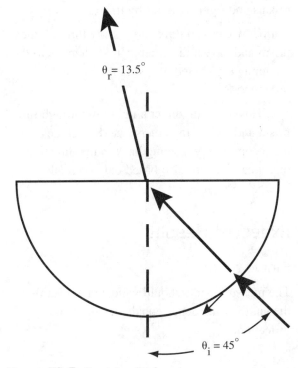

Figure 78-5 *Semicircular lens.*

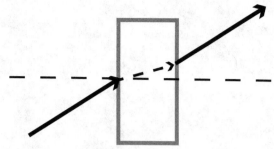

Figure 78-6 *Rectangular prism.*

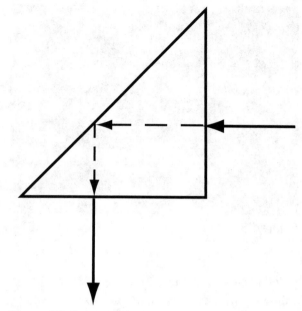

Figure 78-7 *45-degree prism.*

Rectangular prism

Draw a perpendicular line to one of the edges of the prism. (Make sure the edges you are using are clear and not frosted.) Direct the beam toward the point where the perpendicular meets the edge and trace the path of the laser through the prism at various angles.

Right-angle prisms

There are two main types of right-angle prisms: 90°, 45°, 45° and 90°, 30°, 60°.

Here is a challenge. Try it either by working out the light rays first or by just playing with the prisms and figuring it out by trial and error:

a) How can you direct a light ray through the prism and have a ray emerge at 90 degrees to the incoming ray (based only on total internal reflection)?

b) How can you direct a light ray through the prism and have a ray emerge at 180 degrees to the incoming ray heading back in the direction that it came from (also based only on total internal reflection)?

Expected Results

Single mirror

The measurements should validate the idea that the angle of incidence equals the angle of reflection.

Mirrors at a right angle

Regardless of the angle at which the laser ray strikes the mirror, the ray reflected from the second mirror will be parallel to the incoming ray, but heading in the opposite direction.

Convex lens

Rays striking the lens traveling along the centerline will go straight through the lens

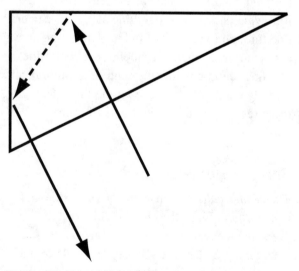

Figure 78-8 *60–30 right-angle prism.*

without being diverted. Rays traveling parallel to the centerline bend toward the centerline and cross at a point (called the focal point) on the opposite side of the lens.

Rays passing through a focal point on the same side of the lens as the laser (same distance as the focal point measured previously) emerge from the lens parallel to the centerline.

Concave lens

No matter where the rays hit the lens, the rays do not cross. Above the centerline, the rays bend up. Below the centerline, the rays bend down.

Semicircular lens

Rays hitting the semicircular side and directed toward the center of the circle bend (or refract) only when emerging from the glass to the air.

Rectangular prism

Rays emerging from the lens are parallel to the incident ray, but offset in the direction that rays move through the lens. As with all these observations, the path the ray takes in the prism is a *straight line*.

Right-angle prism

Rays striking one of the two shorter edges of the (90°, 45°, 45°) prism make a 90-degree turn, and then reflect back.

Rays striking the longer (hypotenuse) edge of the (90°, 30°, 60°) prism make a 180-degree turn, and then reflect back.

60-degree prism

These produce a range of transmitted angles and conditions for total internal reflection.

Why It Works

Lenses are optical devices that refract light passing from one medium (air) through another (glass or plastic), and then typically back to the air. At each interface, the light is bent according to Snell's law.

Mirrors, whether involving one or many surfaces, reflect light in such a way that the angle of incidence equals the angle of refraction.

Other Things to Try

The lenses and mirrors previously described can also be studied using the following additional methods:

- Ray tracing with split beam. You can make or buy an apparatus that projects several parallel beams of light. Basically, the apparatus consists of a bulb covered by an enclosure that shines through parallel openings in the side of the enclosure. The beams of light, when directed at the various lenses and mirrors, show the properties of the devices in a graphic and intuitive way. This approach is better suited to smaller lenses and mirrors.

- Ray tracing by locating images. A more traditional approach is to view a vertical object such as a pin or a nail through the lens. This is accomplished by: a) tracing the outline of the lens, b) locating the position of the image seen through the lens, and c) drawing a line to show the incident, refracted, and transmitted paths of light.

The Point

Lenses and mirrors divert light in predicable ways. Lenses are based on refraction, while mirrors employ reflection. Some lenses are converging and direct rays of light passing through them through a focal point. Other lenses are diverging and direct the rays of light in such a way that they never cross. The reflection at the surface of any plane mirror occurs in such a way that the angle of incidence equals the angle of reflection.

Project 79
Two candles, one flame.

The Idea

No, this is not the title of a bad country western song. This is a great optical illusion that is best shown to a group of observers who have *not* had the benefit of seeing how it was set up beforehand.

What You Need

- 2 nearly identical candles
- match or lighter
- table
- pane of glass (or Plexiglas)
- way to hold the glass perpendicular to the table *safely and securely*. This works with someone simply holding the pane of glass on the table. A ring stand with a beaker clamp (or two) also works well.

Method

Setup

1. Secure the glass perpendicular to the table.
2. Place one candle on the table in the upright position on one side of the glass.
3. Place the other candle upright on the table on the other side of the glass along the same line and the same distance as the first candle.
4. Pick one side of the glass to view.
5. Light the candle on the viewing side.

Figure 79-1 *Photo by S. Grabowski.*

Viewing

1. All observers should be on the side of the pane with the lit candle.
2. Observe the appearance of both candles.

Expected Results

You can see the image of the flame from behind the glass, superimposed on the wick of the (unlit) candle in front of the glass.

Figure 79-2 *Photo by S. Grabowski.*

Why It Works

When light strikes a transparent surface, such as a piece of glass, some of the light is reflected, while some of the light is refracted and transmitted through the glass. The image of the burning candle in front of the glass is reflected. The image of the candle without a flame from behind the glass is transmitted. Both the reflected and transmitted light rays form images that fall on top of each other. This creates the illusion that a single image exists and the candle behind the glass is burning.

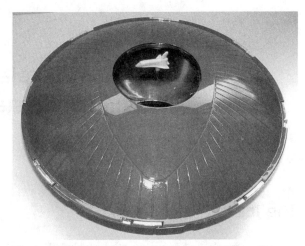

Figure 79-4 *Illusion of the shuttle appears to be floating in the air.*

However, when viewed from a slightly different angle, the real toy shuttle can be seen at the bottom of the dome, producing the virtual image also pictured in Figure 79-5.

This looks so real, it is common for observers to reach in and try to touch the virtual image to convince themselves it is not real.

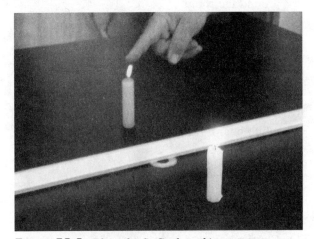

Figure 79-3 *Photo by S. Grabowski.*

The Point

A transparent object, such as glass, can both transmit and reflect light that is incident on it. Reflected light follows the law of reflection, where the angle of incidence equals the angle of reflection.

Other Things to Try

Another way to do this is to replace the unlit candle with a beaker of water. The water level should be above the level of the candle. The reflected image of the lit candle combined with the transmitted image of the beaker of water creates the illusion the candle is burning under water.

A similar illusion can be created by multiple reflections creating a virtual image floating in the air. In Figure 79-4, the image of the space shuttle appears to be hovering over the dome.

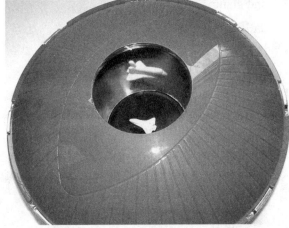

Figure 79-5 *The real toy shuttle seen at the bottom of the dome is the sources of the illusion.*

The Idea

This is a simple and fun way of exploring the law of reflection that provides an initial insight into what is needed to achieve optical alignment. This project (which I first heard from Tom Misniak) makes a good team-building activity and can be used as the basis for a friendly competition.

What You Need

- low-power laser (a laser pointer is fine)

- apparatus (such as a tripod) to mount the laser point and hold it illuminated and stationary for a sustained period (you may need to tape the laser point to keep it on without holding it)

- several plane mirrors

- small white boards

- way to mount the mirrors, such as ring stands and clamps or modeling clay

- dark room

- timer

Method

Competition 1 (Round robin)

1. Caution: Be careful not to shine the laser beam at anyone's eyes. Although the power of the laser should be low, it is a good idea to take care not to expose anyone's eyes.

2. Distribute a mirror and associated mounting hardware to each participant.

3. Draw a target and mount it in a location where everyone has a clear line-of-sight to it.

4. Set up the laser pointer in a central location.

5. Place the participants at different locations around the room.

6. Define a sequence for the beam to reflect from one person to the next and, finally, to the target.

7. Optional: darken the room.

8. Direct the beam to the first mirror. Use the white board, if necessary, to "capture" the beam.

9. Optional: start the timer.

10. Next, align the pointer and the first mirror to make the beam reflect to the second mirror.

11. Continue from one mirror to the next until the last mirror directs the beam to the target, as shown in Figure 80-1.

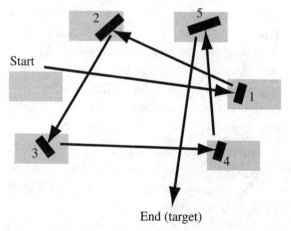

Figure 80-1 *Laser obstacle course.*

12. Optional: in case this gets too easy, require the final beam to pass through a cardboard tube mounted in front of the target.

13. Compasses (for Competition 2).

Competition 2: Back to the target

1. Take similar precautions, distribute mirrors, and establish a target, as in Competition 1.

2. Place the participants at different locations around the room.

3. This time, however, have each participant work out their angles and alignment based on applying the law of reflection and measuring angles.

4. Give each participant a set time.

5. When the time is up, all participants must no longer touch the mirrors. In fact, you can have them leave the area altogether.

6. Then comes the moment of truth, where you see how close each of the participants is able to direct the laser to the target.

Expected Results

It is not unreasonable to have six reflections in about ten minutes. This requires interaction and coordination between groups. One lesson people doing this learn is that small changes at the beginning of the course result in large errors at the end. Even vibrations in the first mirror in the sequence can throw off the alignment downstream. Adjustments may need to be made at each step. Another valuable lesson is there comes a time when it is best to leave the mirror alone and stop making changes.

One other thing that may not immediately be obvious is this is a three-dimensional alignment problem. You not only need to adjust left and right, but also up and down.

Why It Works

This is an application of the law of reflection. Because the reflection angle from one mirror becomes the incident angle of the next mirror, errors in incidence double with each reflection.

Other Things to Try

For the truly dedicated: work out the angles for multiple reflections and back to the target on paper first.

The Point

The angle of incidence equals the angle of reflection. Small alignment errors can be quickly compounded when multiple reflections occur.

Project 81
Light intensity. Putting distance between yourself and a source of light.

The Idea

We all know that stars are intense sources of light, but they appear as faint objects in the sky because they are so far away. How does that work? If you increase the distance between yourself and a light source, does the light become one-half of its original brightness or does it drop off some other way? Check it out and see for yourself.

What You Need

- A light-sensing circuit assembled from
  - solar cell
  - insulated wire (you need two 15-inch lengths with the insulation removed)
  - wire stripper (a pen knife or a pair of scissors will do)
  - ammeter (if you have a multimeter, configure it as an ammeter)
  - soldering iron with (resin core Pb/Sn) solder
  - a third hand (or a second person to help solder)
- or a commercially available light meter, such as shown in Figure 81-1

Figure 81-1 *Light sensor. Courtesy PASCO.*

- light bulb (*not a focused* source, such as a laser or flashlight)
- tape measure
- dark room

Method

Attaching wires to a solar cell to use as a sensor

1. Plug in the soldering iron. Make sure the tip is in a safe place. It gets very hot in a few minutes and should not be in contact with anybody or anything that is flammable.

2. Place the solar cell with the blue side up and locate a pad near the edge intended for wire attachment.

3. As the soldering iron gets hot, "wet" the tip by melting some of the solder on the soldering iron tip.

4. Strip about ¼ inch of the insulation from the end of each of two 15-inch lengths of wire.

5. Position the wire on one of the contact pads on the edge of the solar cell.

6. Position the end of solder from the coil on the pad and near the wire.

7. Touch the soldering iron to the wire to get it as hot as possible. This may happen a bit faster if it is at first raised above the solar cell to avoid heat being conducted by the solar cell. You also want to avoid overheating the solar cell, which could cause it to short out if the heat is excessive. And, you should avoid having any electrical contact—whether it is the wire touching or solder—between the

front pad and the back of the solar cell. This can short out the solar cell and prevent it from generating an electrical output.

8. Touch the solder to the heated wire and, as it melts, have some of the molten solder form a bridge to the solder pad.

9. Remove the soldering iron and don't move until the solder solidifies. If done properly, the solder should stick to both the solar cell pad and the wire forming a mechanical bond.

10. Similarly attach a wire anywhere to the back of the solar cell.

11. Attach the wire coming from the back of the solar cell to the positive terminal of the ammeter. Attach the wire coming from the front of the solar cell to the negative terminal of the ammeter.

12. At this point, you can (carefully) hold the solar cell or mount it by gluing or taping it to a board. Remember, solar cells are extremely fragile and will break if the slightest pressure is put on them. The solar cell may still function if fractured, but reasonable caution can avoid that.

Making the measurement

1. Position the light bulb, so a few meters are in front of it without obstruction.

2. Turn off the room lights.

3. Pick a location such as about 25 cm from the light bulb and take a reading on the light meter. (This starting distance is arbitrary and depends on the sensitivity of the meter you are using.)

4. Record the distance (any choice of unit, inches, or meters can work, but be consistent throughout your investigation). Record the light intensity, as shown in Figure 81-2, in the units in which the light meter is calibrated (such as lumem/m^2 lm/m^2).

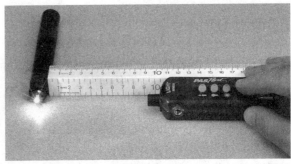

Figure 81-2 *Courtesy PASCO.*

5. If you are using the solar cell, orient it so it is perpendicular to the line between you and the source of light; the unit of measurement will be in amps. Be careful not to block the front of the solar cell with your fingers, which can compromise the accuracy of your reading.

Expected Results

The farther away you get, the less intense the light becomes.

The rate of drop-off is not linear. The farther away you get, the faster the light intensity falls off.

More specifically, the light intensity drops off as the inverse-square of the distance. This is shown graphically in Figure 81-3.

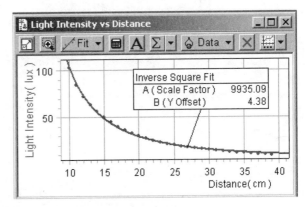

Figure 81-3 *Courtesy PASCO.*

Why It Works

Light intensity is related to the distance from its source according to the equation:

$I = I_o/r^2$

where I represents light intensity at distance, r, between the light source and the point of measurement for an initial intensity, I_o.

Other Things to Try

A similar inverse square law relationship can be found with a source of sound and a sound intensity meter.

The Point

Light intensity drops off as the inverse square of the distance from the source of light.

Project 82

How do we know that light is a wave? Thomas Young's double slit experiment with a diffraction grating.

The Idea

Sir Isaac Newton was definitely no slouch when it came to physics. And, if you asked Newton whether light was a wave or a particle, he would say it was a particle. Newton was actually correct for reasons that would not become clear until several centuries later. Thomas Young proved the opposite was true—that light was (also) a wave. Today, we recognize that light has *both* particle and wave-like behavior. We re-create Young's experiment in this project to explore light's wave-like behavior. Young observed the effect of light emerging from two small slits in an opaque plate. Instead of two slits, you create a similar effect using a diffraction grating which lets you explore the effect of dozens of openings.

What You Need

- diffraction grating available from scientific supply companies, including
  - Edmunds 3001307 13,500 lines/inch http://scientificsonline.com
  - PASCO OS 9127 600 lines/mm (15,000 lines/inch) http://store.pasco.com
  - Frey Scientific 1559099721 15,000 lines/inch http://www.freyscientific.com
- laser pointer
- meterstick
- ruler with metric divisions (centimeters)
- index card
- dark room
- holders to support rulers
- holders for card, diffraction grating, and grating

Method

Setting up

1. Mount the diffraction grating with the lines oriented up and down.

2. Mount the index card, so it is parallel to the diffraction grating.

3. Arrange the meterstick, so you can monitor the distance between the diffraction grating and the screen as you adjust this distance. (If it is convenient to set up, one possible approach is to attach the diffraction grating and screen directly to the meterstick, and then determine the distance between them from the difference in the readings.) A good starting distance is several centimeters.

4. Hold (or secure) the laser pointer, so the laser beam is directed perpendicular to the diffraction grating. See Figure 82-1.

5. Darken the room and turn on the laser. (Caution: As with any project involving a laser, use a low-power laser, such as a laser pointer, and be careful to avoid contact with anyone's eyes.)

6. You should see the path of the laser through the diffraction grating. The brightest spot is called the *central maximum*. Draw a vertical line through the central maximum.

7. On either side of the central maximum, you should also find a much dimmer spot. This is

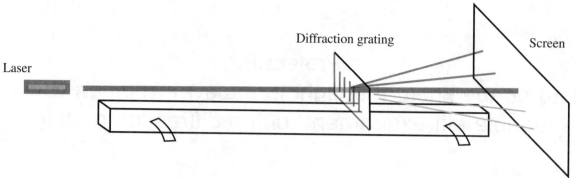

Figure 82-1 *Apparatus used for this project.*

called the *first order line*. (It is more like a spot than a line in our case because we are using light from a laser, rather than the vertical slits originally used by Young.)

Expected Results

A pattern of spots is produced to the right and left of the center line. These are the result of the constructive interference of waves. This proves that light is a wave. Or, more accurately, in addition to having particle-like properties, light is also a wave.

If the distance to the screen is increased, the distance between the bright spots also increases. The distance between the laser and the diffraction grating should not matter, however, because the light strikes the diffraction grating in a perpendicular direction, regardless of how far it is coming from.

Why It Works

When waves meet, if the crests occur at the same time, the waves add. This is called *constructive*

interference. If when waves meet a crest and trough come together, the waves cancel. This is called *destructive interference*.

Interference is a basic characteristic of waves. The light- and dark-spot pattern is the result of interference of the waves emerging from two adjacent openings in the diffraction grating.

Other Things to Try

Once you locate the first order bright spots, you can try to locate the second, third, and possibly higher order lines. This may require a very dark room.

The Point

This project recreates one of the most significant experiments of the twentieth century in which Thomas Young demonstrated that light is a wave. Interference patterns are a unique characteristic of waves. Because light in this experiment exhibits an interference pattern, it proves conclusively that light is a wave.

Project 83
How to measure the size of a light wave.

The Idea

Light is a wave. The interference pattern created by overlapping light waves is different for different wavelengths of light. We can use this to determine this wavelength.

What You Need

- diffraction grating of known separation between the lines
- laser pointer of known wavelength (low-cost red lasers are typically 650–670 nm, green lasers are in the range of 535 nm)
- apparatus used in the previous project (including a meterstick, ruler, index card, and associated clamps and holders)
- protractor
- dark room

Method

1. Set up the apparatus used in Project 81.
2. Determine the spacing *between* the lines of the diffraction grating. Diffraction grating suppliers typically identify the number of lines in a given distance. For instance, if a diffraction grating has 15,000 lines per inch, the spacing from the center-to-center of the rule lines is 1 inch/ 15,000 lines or 0.000067 inches between lines or 0.0026 m between lines.
3. Darken the room. (Caution: As with any project involving a laser, use a low-power

laser, such as a laser pointer, and be careful to avoid contact with anyone's eyes.)

4. On either side of the central maximum, you should also find a much dimmer spot. This is called the *first order maximum*. With a pencil, mark the distance to each of the first order maxima.
5. Using the meterstick and protractor, measure the angle between the point where the laser light passes through the diffraction grating and the first order spot on the card. Measure both the angle to the right and to the left of the centerline. The angle to the left and to the right of the center spot should be very close to each other. Repeating this measurement and taking the average can give a more accurate value.
6. If the room is dark enough and everything else is working, it is likely you will be able to see the second and, possibly, the third-order maxima. Note the distance and save the data for the Other Things to Try section.
7. Repeat this using several combinations of screen distance.
8. The diffraction grating equation is

$$\lambda = \frac{d\sin\theta}{n}$$

where λ (Greek letter "lambda") is the wavelength (in meters), *n* is the order of the band = 1, 2, 3 … For the first order, n = 1, and *d* is the distance (in meters) between lines on the diffraction grating.

9. Use this data table to organize your data. Convert all your data to meters. If your

	1st trial	2nd trial	3rd trial	4th trial
L (in m) distance from grating to screen				
Angle (in degrees) to first order bright spot (on right)				
Angle (in degrees) to first order bright spot (on left)				
Average angle (in degrees) to first order bright spot				
d (in m) distance between grating lines				
λ (in m) wavelength of light				Average:

measurements are in centimeters, convert them to meters by dividing by 1000.

10. Compare the average wavelength you find to the expected wavelength.

11. Note, if you really want to nail this measurement, remember the lines themselves take up some of that distance. The distance between the lines (which is what we really need) is slightly less than the center-to-center distance between lines. You can calibrate for this discrepancy by using the previous diffraction grating with a light source with a known wavelength. By measuring the distance between card and grating, the location of the first order maxima, and using the known wavelength, you can solve for d, the spacing *between* the lines. Although this gets you into a bit of a chicken and egg situation, it does enable the most accurate results.

Expected Results

For commonly used diffraction gratings, the expected wavelength can be found by measuring a bright light maximum at the following angle.

Laser color	Laser wavelength	Laser wavelength	Angle
Red	660 nm	6.60×10^{-6} m	21.2°
Reddish orange	635 nm	6.35×10^{-6} m	20.5°
Green	535 nm	5.35×10^{-6} m	17.5°

Why It Works

The distance between the bright spots established when light of a given frequency passes through a diffraction grating can be used to measure the wavelength of a source of light. If the spacing of the grating and the distance to the image are known, the wavelength can be determined.

Other Things to Try

For simplicity, we only used the first order maximum. You can try the previous measurement also using the second and, possibly, the third order maxima. This technique should give the same wavelength, regardless of which order is being used, so finding the same or a similar result from different orders can give you a confirmation that you are doing something right.

The spectral breakdown of light emitted by a hydrogen atom can also be detected using a high-sensitivity light sensor, such as PASCO part number PS-2176. An example of this is shown in Project 120.

The optical tracks on a CD or the grooves on a vinyl recording are effectively diffraction gratings. Use the diffraction grating equation to find out the spacing of the tracks on a CD or record. Basically, the reflection from the closely spaced tracks on a CD can give a similar interference pattern, such as that produced by transmission through a diffraction grating. Details can be found in an article called "Using a Laser

Pointer to Measure the Data Track Spacing on CDs and DVDs" (www.sciencebuddies.org/science-fair-projects/project_ideas/Phys_p011.shtml?from…).

The Point

A *diffraction grating* is a device that produces an interference pattern when a light is shined on it.

If that light has one frequency (or is monochromatic), such as a laser, the interference pattern can tell us what the frequency is. This capability forms the basis of many analytical techniques that require measuring the frequency of light with great precision.

Project 84
The speed of light in your kitchen.
Visiting the local hot spots.

The Idea

Light is an electromagnetic wave. Because of its extremely high speed, it is difficult to measure the speed of light directly. Historically, astronomers such as O. Roemer used distances on the scale of planetary orbits to get a handle on how fast light traveled through space. In this project, you measure the speed of light right in your own kitchen. The technique is similar in principle to the approach we used in Project 73 where we found the speed of sound by finding its resonant wavelength. Here, you use a microwave oven to establish a standing wave that can be used to *estimate* the speed of light.

What You Need

- microwave oven
- sheets of sliced cheese, bars of chocolate, or about five eggs
- sheet or plate to hold above food items without rotating in the oven—some ideas include a rectangular wood or plastic cutting board, a rectangular Pyrex baking dish, a round dish that fits as close wall-to-wall as possible, or a sheet of poster board cut to size. Obviously, remember *no metal* should go in the microwave oven.
- ruler
- calculator
- light, so you can see inside the oven (the oven may have an adequate light built in)

Method

1. Microwave ovens rotate to spread out the hotspots in the oven. In this experiment, we want to detect these hot spots. So, if your microwave oven has a rotating tray, remove it from your microwave oven.

2. Put a nonrotating sheet or plate in the bottom of the microwave oven.

3. Cover the plate with the microwavable food: cheese slices or chocolate slabs. The layer should be as uniform as possible in thickness and composition. If you choose to use egg whites, pour a thin layer into a suitable dish and spread it out to form a thin, uniform layer.

4. Look through the glass window of the microwave oven and start the microwave oven on the lowest available power setting. Use a light shining from outside if that helps you observe what is going on inside the oven. If you don't have a window, you need to cook in increments. Because microwaves differ so much in power, you need to determine an appropriate amount of time to use for this: 10 to 15 seconds is a good place to start.

5. Continue running the microwave oven until you notice the first signs of melt spots or cooking.

6. Stop the microwave oven and identify the pattern of melt spots. Unless you are sure the microwave oven has run long enough to establish a clear melt spot pattern, do not move the tray in the oven yet.

7. Measure the center-to-center distance between adjacent melts spots. An example for what you are looking for is shown in Figure 84-1.

Figure 84-1 *Microwave standing wave nodes are indicated by hot spots in melted cheese.*

8. Because one-half of a wavelength fits between each hot spot, the wavelengths for the microwaves in the oven are twice the center-to-center distance measured.

9. Look for a label or search online for the specific microwave frequency used in your microwave oven. If you cannot easily find this frequency, you can use 2450 MHz, which is the frequency at which most commercial microwave ovens operate.

10. When you finish, you can make grilled cheese sandwiches, s'mores, or egg-white omelets with the leftover food ingredients.

11. Calculate the speed of light using the equation:

    $c = \lambda f$

    or speed of light = the wavelength/frequency

    For λ use the wavelength (in meters) from twice the center-to-center hot spot distance.

    For f, use 2450 MHz, which is 2,450,000,000 Hz or 2.45×10^9 Hz (unless otherwise indicated on the microwave oven).

Expected Results

Suppose you find the average of the hot-spot distance is 6 cm.

The wavelength of the microwave resonant in the microwave oven is 12 cm.

In meters, this is 12 cm × 0.01 m/cm = 0.12 m.

The speed of light is then, c = 0.12m × 2,450,000,000Hz = 2.94×10^8 m/s.

This is reasonably close to the accepted value, which is just under 3×10^8 m/s (300,000,000 m/s). Remember, microwaves may vary in how they create standing waves and an error factor is associated with heat distribution on the heat surface. For this reason, this measurement can be only expected to give ballpark, not precisely accurate, values.

Why It Works

A microwave oven produces a resonant wave in the oven chamber similar to that of a vibrating guitar string. The hot spots are like the nodes or the points where the ends of the string are held. A complete wave is a cycle up and down, so only a half wave fits between the two nodes in both cases. From knowing the frequency of the microwaves and measuring its wavelength, we can find the speed of light.

Other Things to Try

A more sophisticated, but more precise way to measure the speed of light is to detect the interference between light waves separated by a measurable distance. Equipment to do this is available from scientific supply vendors, such as PASCO.

The Point

The speed of a wave, such as light, can be determined from its wavelength and frequency. The wavelength of a microwave oven can be found by the distance between the nodes of a resonant standing wave.

Project 85
Refraction. How fast does light travel in air or water?

The Idea

Light travels at its top speed in a vacuum and at (nearly) its top speed in air. When light moves through other transparent materials, it slows down. If it goes from one material to another at an angle, the light will bend. The more it slows down, the more it bends. In this project you will compare how much light bends in various materials. This bending is called *refraction* and it gives us a way to determine how fast light travels in a transparent material.

What You Need

- square or rectangular piece of glass about ¼ inch thick and a few inches in length and width (at least two opposing sides must be clear)
- laser pointer
- semicircular plastic container filled with water
- protractor

Method

Laser

1. Place the piece of glass on the paper.
2. Trace the shape of the glass.
3. Darken the room.
4. Put a dot on one side of the glass to provide a target for the laser.

5. Draw a line perpendicular to the edge of the glass at that point. Extend the line, so it extends under the glass, as well as going into it.
6. Place the laser, so its beam forms an angle with respect to the perpendicular line you just drew. Mark the position of the laser.
7. Darken the room.
8. Shine the laser and direct its beam toward the target dot. Angle the beam vertically, so you can see its path both before entering and after exiting the glass. (It is OK if you don't see the entry and exit beams at the same time.)
9. Place a dot where you see the laser beam exit from the glass and one or two dots along its path.
10. Connect the dot where the light strikes the glass with the point where the light ray emerges from the glass back into the air.
11. Measure the angles that:
    - the incoming ray made with the perpendicular line (θ^i).
    - the ray going through the glass made with the edge where the light entered the glass (θ_r).
12. Try this with other transparent materials such as water (in a plastic case).

Expected Results

Going from air into water, the light path is bent to give a smaller angle with respect to the perpendicular line. Specific results are given in the following chart:

Incident angle coming from air	Refracted angle into glass (n=1.45)	Refracted angle into water (n=1.33)
10	6.9	7.5
20	13.6	14.9
30	20.2	22.1
40	26.3	28.9
50	31.9	35.2
60	36.7	40.6
70	40.4	45.0
80	42.8	47.8

Why It Works

The relationship between the incident and refracted angles is given by Snell's law, which states:

$$n_i \sin(\theta_i) = n_r \sin(\theta_r)$$

where n_i is the index of refraction where the ray is incident and n_r is the index of refraction where the ray is refracted. Both are measures of the speed of light in the various materials, The index of refraction for any material is given by $n = c/v$. The physical arrangement for this is shown in Figure 85-1.

The index of refraction for air, which is 1.0, indicates that light is traveling at its maximum speed.

Because light slows down in glass, *n* for glass is 1.45.

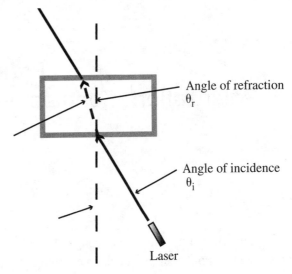

Figure 85-1

Other Things to Try

Find the index of refraction for the glass or water using the equation:

$$n_i \sin(\theta_i) = n_r \sin(\theta_r)$$

using n_i for air = 1.0 and θ_I and θ_r as defined in the previous step to find the index of refraction in the material of interest. The index of refraction is given by:

$$n = c/v$$

The index of refraction is a measure of the speed of light in a particular material, v, compared to the speed of light in a vacuum which is given by $c = 3.0 \times 10^8$ m/s.

The speed of light in a particular medium is given by $v = c/nr$.

The Point

Light travels at a slower velocity when it goes through materials other than vacuum or air. When light hits a boundary going from a material where the light is faster to one where it is slower, the light bends toward the perpendicular line.

Project 86
Polarization. Sunglasses and calculator displays.

The Idea

The orientation of the crests and troughs of light waves can be horizontal, vertical, or anything in between. *Unpolarized light* consists of a random mix of orientations. *Polarized light* has only one direction. This gives it the unique properties used in liquid crystal displays found in many television screens and computer monitors. Sunglasses reduce glare by allowing only selected orientations of polarized light through. This experiment explores how to identify whether light is polarized and how the transmission of polarized light can be controlled.

What You Need

- 2 polarized sheets
- calculator or other LCD display, such as a laptop computer
- polarized sunglasses
- light source
- shallow tray
- water
- few rocks
- sheet of glass
- optional: protractor, light sensor

Method

Transmission through polarized sheets

1. Take one of the polarized sheets. Hold it in front of a light source (a lamp or an open window) and rotate it. Turn the sheet a full 360 degrees, holding the sheet so it remains roughly perpendicular to your field of view.

2. Try this with the other sheet.

3. Now, with both sheets, hold them in front of the light, one in front of the other, and rotate only one of them. Observe what happens. Add in other combinations: rotate both in the same direction, rotate both, but in different directions. Describe the effect of the sheets. Is the light from your light source polarized?

4. Using a nondestructive method, such as applying a small piece of tape, identify an edge on each, which when placed together, blocks the maximum amount of light.

5. Note: This is good to do using an overhead projector or the light from an LCD projector.

Reflections

1. Place a small flat mirror face up on a table.

2. Place a light source on one side of the mirror.

3. View the reflected light through one of the polarized sheets as you rotate the sheet. Is that light polarized? View through a range of reflected angles from nearly perpendicular to a very glancing angle to the mirror.

4. Place the two sheets, one on top of the other, but with taped edges aligned, so both sheets have the same polarization plane. What happens when you rotate the sheets, both with respect to the mirror and to each other?

5. Repeat 1–4, using light reflected from a square of glass.

Laptop/sunglasses

1. Hold a polarized sheet in front of the LCD (liquid crystal display) of a laptop computer or digital calculator.

2. Rotate the polarized sheet. What can you conclude about the LCD of the laptop?

3. Take the two lenses from an old (no longer needed) pair of polarized sunglasses. Hold them—one in front of the other—and view a light source. Are the glasses polarized? You also can try this with two pairs of polarized sunglasses.

Expected Results

Light will pass through polarized sheets with little loss when the directions of polarization for both sheets line up. As the sheets are rotated, more and more of the light is blocked. With the direction of polarization of the two sheets at right angles, almost no light can pass through, as shown in Figure 86-1. This can be quantified in

Malus's law, which is addressed later in this section.

A light meter is a good way to quantify the amount of light passing through a filter. Figure 86-2 provides an approximate visual reference to evaluate the amount of light transmission through a set of polarizing filters.

Reflected light can be polarized. This can be determined by observing the effect of a single polarized filter on reflected light.

The light from a LCD, such as a laptop screen, is also polarized. This can be seen by rotating a polarized filter, such as polarized sunglasses, in front of an LCD screen, as shown in Figure 86-3 and Figure 86-4.

Why It Works

Light is an electromagnetic wave that propagates along a line. If we can imagine looking down that line, we would see the waves from most light sources moving up and down in any direction. For unpolarized light, the wave oscillations are randomly distributed over 360 degrees. A polarized filter selects only one of the polarization planes. Reflection polarizes light by favoring light in the plane of the reflecting surface.

Figure 86-1

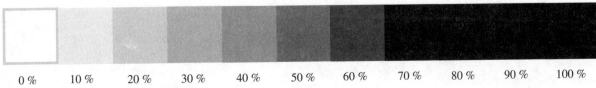

| 0 % | 10 % | 20 % | 30 % | 40 % | 50 % | 60 % | 70 % | 80 % | 90 % | 100 % |

Figure 86-2 *Percentage of light blocked by polarizing filters.*

225

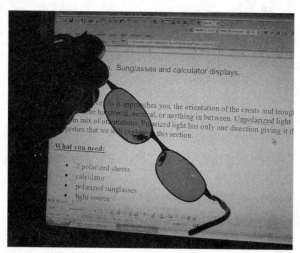

Figure 86-3 *Polarized filter (sunglasses) oriented to pass polarized light from laptop.*

Figure 86-4 *Polarized filter (sunglasses) oriented to block polarized light from laptop.*

Other Things to Try

Malus's law

We saw how the more we rotated the polarized sheets, the less light they transmitted. Here, we find the physical model for that effect.

1. Start with the two polarized sheets oriented (with taped sides aligned) to allow the maximum light to be transmitted.

2. Find a way to measure or estimate the amount of light transmitted. Depending on your resources, this can include:

   - Estimating the transmission visually by using a chart ranging from white 100 percent transmission to black 0 percent transmission. Figure 86-5 may be a rough guide.

   - Comparing the transmission using a set of calibrated neutral-density filters (that may be available in some labs). These transmit a specific amount of light without changing the color, so they might serve as a good visual comparison.

   - Using a light meter to measure the light transmitted. The percent would be the ratio of the light transmitted at a particular misorientation angle divided by the amount of light transmitted through the polarized filters when they are aligned. This works best if the room is darkened and the light from the source is isolated from the meter. One way to do this is to put the light in a box and cut a square hole smaller than the filters. The light meter can be purchased from a science supply company or made from a solar cell with solder wire leads connected to an ammeter (as described in Project 81).

3. Measure or estimate the amount of light transmission as a function of misalignment angle of the polarized sheets. Use the following data table to organize your data.

Misalignment angle, θ (in degrees)	0	10	20	30	40	50	60	70	80	90
Fraction of light transmitted										

The expected results are described by this is called Malus's law.

The greater the misorientation, the less light is transmitted. The drop-off, however, is not linear but is, instead, given by:

Fraction of light transmitted, $I/I_o = \cos^2\theta$

The expected results are given in Figure 86-5.

Find Brewster's angle

Previously in this section, you saw that light reflecting from a piece of glass can be polarized if the angle (with respect to the normal) is great enough. That angle is called *Brewster's angle*.

1. Place a sheet of glass flat on a table.

2. Using a protractor as a guide, view the reflected light at various angles. Use a straight edge positioned near the protractor as a visual guide to establish the reflected angle.

3. Determine the maximum angle with respect to the perpendicular that results in reflected polarized light. That is Brewster's angle.

4. Compare your result with the expected value of Brewster's angle given by:

$n = \tan\theta_p$

where n is the index of refraction for the glass and θ_p is the angle where the reflected light is completely polarized.

Some typical values are given in the following table:

Material	Index of refraction	Brewster's angle, θ_p
Water	1.33	53.1°
Crown glass	1.52	56.7°
Flint glass	1.45–2.00	55–63°

Finding an object under water

1. Place several objects (coins, rocks) in a pan of water.

2. Cover the objects with several inches of water.

3. Establish a light source at an angle.

4. Find a position to view the surface of the water, so you see the reflected light shimmering at the surface of the water obscuring the objects below.

5. View the light using the polarizing filter. View at angles both greater and lesser than Brewster's angle through the polarizing filter. Under what conditions are you able to see the underwater objects?

The Point

Typical light sources, such as light bulbs or the sun, produce unpolarized light that has electromagnetic waves oriented randomly. Light can be polarized by filters or certain reflecting surfaces that select a specific orientation of the light waves.

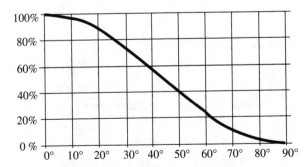

Figure 86-5 *Malus's law.*

Project 87

What is the wire of a fiber-optic network? Total internal reflection using a laser and a tank of water.

The Idea

Much of the digital communication that circulates around the world on the Internet consists of light traveling thousands of miles through glass threads thinner than human hair. Total internal reflection is the physical process that keeps these information pulses confined within these tiny optical fibers. In this experiment, you can set up various ways to explore how transparent materials guide light by total internal reflection.

What You Need

- laser pointer
- fish tank (preferably with a glass bottom)
- tablespoon of milk
- dark room
- sink
- one 2-L clear plastic soda bottle

Method

Total internal reflection from the surface of water

1. Fill the tank with water.
2. Turn off the room lights.
3. Direct the laser from the side of the tank under the level of the water toward the surface.
4. Add enough milk to the water in the tank so the path of the laser is visible as it passes through the water.

5. If the fish tank has a glass bottom, direct the laser through the fish tank from the bottom at various angles.
6. Determine the maximum angle that will allow the light to emerge from the tank.

A liquid light pipe

1. Punch a small hole about 1 mm in diameter in the side of the plastic gallon container.
2. Fill the container with water.
3. Darken the room.
4. As the water flows from the hole into the sink, shine the laser from the other side of the bottle, but aim it at the hole.
5. Observe how the light passes through the bottle and is guided through the water stream pouring out of the bottle. This is because the light is trapped in the water stream by total internal reflection. With an index of refraction of 1.33 for water, any light incident on the surface at an angle greater than 48.7 degrees will be totally reflected. This is much lower than the angles the light makes with the water stream.

Acrylic cylinder lightguide

An acrylic (or other transparent material) makes a good demonstration of light trapped in a medium by total internal reflection. Send the beam through one of the flat ends. Regardless of what angle you direct the beam, the light will not refract out through the sides of the cylinder.

Observing light fibers

Optical fibers show very convincingly how light travels through a thin transparent material. Optical fibers are available from scientific supply companies and novelty lighting stores.

Expected Results

Light passing through the water and striking the surface passes through to the air if the angle is above the critical angle for water.

Figure 87-1 shows light passing through plastic at an angle less than the critical angle. Notice that most of the light is refracted (or goes through) the lens with a smaller part of the light being reflected at the flat surface.

However, once the incident angle coming from the plastic back into the air is greater than the critical angle for those materials, the light experiences total internal reflection, as shown in Figure 87-2. Here, light is effectively trapped inside the plastic and cannot emerge back into the air.

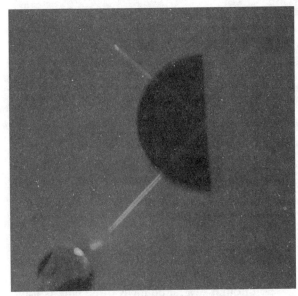

Figure 87-2 *Total internal reflection of incident light.*

Why It Works

When light enters a material where it goes faster, the angle approaches 90 degrees as the incident angle increases. Because the angle of refraction cannot be greater than 90 degrees, light above the critical angle is totally reflected.

Other Things to Try

Variable index of refraction. (Based on a lecture demonstration posted at the UCB web site: www.mip.berkeley.edu/physics/E+60+40.html.)

Refraction takes place when light goes from one material to another. Fiber-optic cable and other optical devices take advantage of a variable index of refraction to guide the light in a channel. To do this:

1. Cover the bottom of an empty fish tank evenly with about 1 cm (about ½ inch) of granular sugar.

2. Add water to the tank as slowly and carefully as you can, so you disturb the layer of the sugar as little as possible. Warm water at about 70 degrees C will enable the sugar to dissolve more quickly.

Figure 87-1 *Incident light is refracted and reflected.*

3. Fill the tank with a few inches or so of water.

4. Let the tank remain undisturbed for a few days, allowing the sugar to slowly dissolve. The concentration of the sugar solution—and, as a result, the index of refraction—will vary with the height above the bottom of the tank.

5. Darken the room.

6. Direct a laser from the side of the tank at an angle with respect to the bottom of the tank. The beam should be guided, as if through an invisible light pipe in the tank. This use of a variable index of refraction is similar to the way a fiber-optic cable guides a light beam. If the laser is directed at an angle, the light may go through several bounces if the tank is long enough and the distribution of the sugar is uniform. This also illustrates how mirages form. When light passes through cooler air to warmer air over a road surface, the variable index of refraction guides the light in a way that creates the illusion of water lying on the road.

Light guides

Optical fibers can be found in toy and novelty stores, and as a part of certain lamps. These work on the principle of total internal reflection and they show how light can be "piped" around corners. A similar way to explore this is to shine a laser into an acrylic cylinder. Although a good bit of scattering is within the acrylic, the light is trapped inside in a similar manner to an optical fiber, as shown in Figure 87-3.

Figure 87-3 *Light coming in one side of the tube is trapped by total internal reflection and emerges from the other side.*

The Point

If the angle that a light ray strikes a surface is too large, the light will not pass through to the other medium. This can only happen if the speed of light is faster in the second medium (or if the index of refraction in the second medium is smaller). For angles greater than that critical angle, no light is refracted, but is instead totally reflected back into the material from which it originated.

Project 88
The disappearing beaker.

The Idea

This demonstration lets you set up a cloaking shield that makes a glass object disappear.

What You Need

- Pyrex beaker
- other glass objects such as Pyrex stirring rods and magnifying glasses
- transparent container large enough to hold the beaker
- cooking oil (such as Wesson, baby oil, Karo syrup, or light and heavy mineral oil)

Figure 88-1

Method

1. Place the beaker in the larger container.
2. Fill the container with the oil.
3. Immerse the beaker in the oil and slowly pour the oil into the beaker.
4. Observe what happens when other glass objects are placed in the oil (Figure 88-1).

Expected Results

As the oil level rises above the beaker, the glass can no longer be seen. If any markings are on the side of the beaker, they will still be noticeable, as seen in Figure 88-2.

If you use other liquids, similar results may be obtained, but you may have to do some fine tuning. Other types of glass and other oils

Figure 88-2 *Disappearing beaker.*

(including the "lite" version of cooking oils) may be less perfect, leaving some ghost images that are less noticeable the farther away your "audience" is. A mixture of a heavy and light mineral oil (in a ratio of about 2:1 to start) should match Pyrex and be adjustable for other types of glass. Karo syrup is a close match to Pyrex and can be diluted with water to match other types of glass.

The magnifying glass will not enlarge images when submerged.

The precise index of refraction for these materials may vary slightly with temperature. Also, imperfections in the glass may make it difficult to make the very last trace disappear on some samples.

Why It Works

Objects are visible to the extent that they are able to reflect light. If an object immersed in a liquid has an index of refraction that is different than the object, some of the light is refracted through the object and some is reflected back to the observer. However, if the object has exactly the same index of refraction as the immersed object, the light will neither reflect nor refract at the interface between the object and the liquid it is immersed in. In that case, the object will appear to be invisible.

Lenses, such as magnifying glasses, work by virtue of their index of refraction being *different* than the index of the medium around it. If the index of refraction surrounding the lens is increased from 1.0, which is the index of refraction of air, to an index very close to glass, the light rays will *not* be bent through a focal point and magnification will not occur.

Both Wesson cooking oil and Pyrex glass have a nearly identical index of refraction of about n = 1.474, making them particularly well matched for this demonstration.

Other Things to Try

A similar experiment along these lines is first to pour a water-alcohol mixture to a beaker, and then cover it with cooking oil. If a small enough amount of alcohol is in the mixture (less than twice as much oil as water), the oil will float and it will also not mix with the alcohol-water layer. If you add a few drops of food coloring to the water-alcohol mixture, the effect is easier to see. Viewing from the top, the water-alcohol layer is invisible. This is the result of total internal reflection at the interface between the layers.

The Point

Transparent objects are visible because of reflection from their surface. Transparent objects will partially reflect and partially refract light if there is a difference between their index of refraction and that of their surroundings. If the index of the material and their surroundings is the same, the object will appear to be invisible.

Hot and Cold

Project 89
How much heat is needed to melt Greenland?
Heat of fusion.

The Idea

Ice anywhere requires a certain specific amount of heat to melt. Ice melts at 0°C. Once at that *temperature*, the amount of *heat* needed depends only on how much ice you have. Heat can either change the temperature of something or cause it to change from one state to another. In this experiment, that change is from solid to liquid. You will determine how much heat is needed to melt a given mass of ice by carefully keeping track of temperature changes and heat flows.

What You Need

- Styrofoam cup
- cube of ice
- graduated cylinder (250 mL)
- water
- thermometer
- stirring rod
- scale

Method

1. Fill the beaker with exactly 150 mL of water at a temperature of at least 25 degrees centigrade. This results in a mass of the water, m_w, of 150 g.

2. Remove an ice cube from the freezer and let it sit out until it just begins to melt. This establishes its temperature at (very close to) 0 degrees centigrade.

3. Measure the mass of the ice cube, m_{ice}, in grams. If significant melting has occurred, you can use a paper towel to absorb any excess liquid water before measuring the mass.

4. Measure the initial temperature of the water, T_i, before the ice cube is added.

5. Drop in the ice cube. Stir gently. Measure the final temperature, T_f, as soon as the ice cube has completely melted.

6. Calculate how much heat was needed to melt the ice through the following steps (see Figure 89-1):

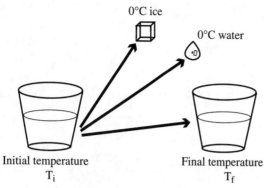

0°C ice

0°C water

Initial temperature
T_i

Final temperature
T_f

Figure 89-1 *Heat is transferred* from the water *to raise the temperature of the ice and water and to melt the ice.*

- Heat extracted from the water

  Given by $Q_w = m_w C_w (T_f - T_i)$

  where C_w is the specific heat of water = 4.18J/g°C. This means that 4.18 joules of energy (which is how energy is measured) are needed to raise each gram of water every 1 degree centigrade.

- Heat needed to bring the melted ice from 0°C to the final liquid temperature

  Given by $Q_{melted\ ice} = m_{ice} C_w (T_f - 0)$

- Heat needed to melt 1 gram of ice

  $H_f = (Q_w - Q_{melted\ ice}) / m_{ice}$

Expected Results

The expected result is $H_f = 334$ J/g. This means 334 joules of heat energy are needed to melt 1 gram of ice.

Why It Works

When a cube of ice is placed in a beaker of water, some heat is taken from the water. This loss of heat results in the water being brought to a lower equilibrium temperature, which is lower than the starting temperature. The heat lost by the water accomplishes two things: 1) it melts the ice, and

2) it brings the liquid resulting from the melting ice up to the equilibrium temperature. This gives the overall equation:

$Q_w = Q_{melted\ ice} + m_{ice}H_f$

Solving for H_f gives the equation used to find the heat of fusion for ice.

Other Things to Try

Determine the temperature of a freezer. As before, place an ice cube (this time *immediately* removed from the freezer) in a Styrofoam cup filled with water of a known temperature. Do this by using the known value for the heat of fusion of ice and the heat transfer equation used previously.

Determine the temperature of a hot object (such as a red-hot nail) by measuring the temperature before and the equilibrium temperature after immersion of the object. Then solve the heat equations for the initial temperature of the object.

Calculate the amount of heat needed to melt the Greenland ice caps: Greenland contains 2.85 million cubic kilometers of ice or 1.4×10^{14} ice cubes each with a mass of 20 g. Based on the heat of fusion for ice (0 degrees C) determined in this experiment, the Greenland ice caps would require 9.5×10^{23} Joules of heat energy to melt.

The Point

When objects at different temperatures are mixed, they result in an equilibrium temperature that is between the highest and lowest temperatures in the mixture. The amount of heat needed to cause a certain temperature change for a given mass of material is characterized by the specific heat for that material. In addition to causing temperature change in a material, some heat (called *latent heat*) results in a change of phase from solid to liquid.

Project 90
A water thermometer.

The Idea

All materials (with the one notable exception of ice at around 4°C) expand when heated. It is hard to notice the difference in volume between a hot cup of tea and the volume in the cup when the tea cools. This experiment gives a way to intensify the effect of the thermal expansion, so it can be measured and compared to a known value.

What You Need

- 250 mL flask
- 2-hole rubber stopper that fits in the flask
- hotplate
- 1000 mL beaker (or one large enough for you to place the flask in)
- glass tube that fits through the rubber stopper (about 15 inches long)
- water
- glass thermometer that fits through the second hole of the stopper (if you don't have a thermometer that works, then you need a one-hole stopper)
- Vaseline (or glycerin) and a towel to help slide the glass and the thermometer into the stopper
- ruler
- ring stand with a beaker or test tube clamp

Method

1. Determine the radius of the glass tube by:
   - Partially filling the glass tube with water. Place your finger over the top of the tube to keep the liquid from sliding out while you are making this measurement.
   - Measure the height, h_o, of the water column.
   - Release the volume into the graduated cylinder and measure the volume.
   - The volume of the liquid measured $V = \pi h r2$.
   - The radius of the tube is $r = (V/\pi h_o)^{\frac{1}{2}}$.

2. Carefully slide the glass tube into the stopper. Use a little Vaseline or glycerin as a lubricant and protect your hands with a towel as you push. *Don't force it.* A few inches of the tube should extend below the stopper with the rest sticking out above.

3. Slide the thermometer into the other hole, so the bottom of the thermometer is positioned near the center of the flask.

4. Completely fill the flask with water. (Add food coloring if you like—your choice of color.)

5. Insert the stopper. A small amount of liquid may spill out over the side of the flask and some may be forced up the tube.

6. Place the flask with the stopper in the beaker.

7. Fill the beaker with water to cover the flask.

8. Place the beaker on the hot plate.

9. Turn on the hotplate. The apparatus for this experiment is pictured in Figure 90-1.

10. Record the temperature in the flask and note the position of the liquid in the glass tube. (If you have a Sharpie handy, mark it on the glass.)

11. When the temperature in the flask rises a few degrees, record the temperature and measure the increase in the height of the liquid in the flask.

12. The *increase in volume* for a given temperature increase is given by $V = \pi h r^2$, where h is the measured height (in meters) and r is the inner radius of the glass previously. In setting this up, some liquid likely will extend into the tube at your starting temperature. If this is the case, define this as your zero point and *take h as the distance the liquid rises into the tube*. (The small difference in volume resulting from the liquid that initially rises into the tube is not significant for this measurement, but if you are very picky, you can correct for this on principle.)

Expected Results

A given volume of water expands by a factor that is 0.000207 (or 2.07×10^{-4}) of its original volume for every 1 degree increase centigrade. This volume is distributed between the flask and the tube.

Why It Works

Nearly all materials expand when they are heated. The amount of expansion is characterized by something called the *coefficient of expansion*. In the case of solids, the expansion in one direction is called the coefficient of *linear* expansion. Multiplying the original length by the coefficient of linear expansion gives how much longer the object is.

Volume works almost the same way, except in the three dimensions. The *coefficient of volume expansion* indicates how much volume is added to a (solid or liquid) material for every degree the temperature increases.

Other Things to Try

Design and calibrate a water thermometer using the coefficient of volume expansion for water and the dimensions you determined for the glass tube.

The Point

The amount a material expands when heated is called the *coefficient of volume expansion*. We constrained the expansion of a larger volume of water in the flask to primarily one dimension in the tube. This magnified the effect of the expansion, so we were able to measure it.

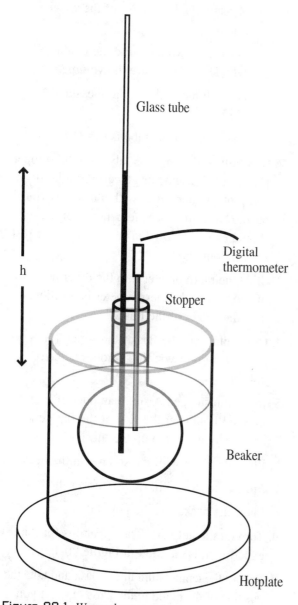

Figure 90-1 *Water thermometer.*

Glass tube

h

Digital thermometer

Stopper

Beaker

Hotplate

Project 91
What is the coldest possible temperature?
Estimating absolute zero.

The Idea

What is the coldest thing possible? Some people might say it is giving a physics test on a Friday afternoon before winter break. (Now that's cold!) But, the coldest temperature possible is absolute zero. This experiment is a nice, simple way to estimate this fundamental property of nature. With some extra care, a more accurate value can be obtained. We know that matter contracts as it gets cold. The basic idea here is to determine what would be the temperature if the volume were to contract to the point where it approached zero. We can't get to that point. In fact, we can't even get close in an ordinary lab. But, we can measure how much the volume changes for a given change in temperature and make a graph to determine at what temperature the volume would be zero. That temperature is *absolute zero*.

What You Need

To estimate absolute zero

- 250 mL Pyrex flask
- pair of tongs suitable for safely handling a hot flask
- beaker (large enough to fully immerse the flask)
- hotplate
- thermometer
- bucket
- graduated cylinder

To measure absolute zero with greater precision

- A temperature volume or pressure-volume apparatus, such as shown in Figure 91-1.

Method

1. Put the beaker on the hotplate.
2. Fill the beaker with water to a level that will allow the flask to be immersed without causing the water to overflow.
3. Turn on the hotplate.
4. Place the empty flask in the beaker, so it is heated from outside, but without having the heated water spill into the flask.
5. After the air in the flask has had a chance to reach equilibrium with the heated water (about 5–10 minutes at a constant temperature), measure the temperature of the water. The boiling point is a good stable measurement point, but a temperature less than this can be used if it is stable.
6. Fill the bucket with cold water. You can use ice to bring the temperature down.

Figure 91-1 *Courtesy PASCO.*

7. Remove the flask and immediately place it neck side down in the bucket. Hold the neck of the flask underwater as it cools. You may find it helpful to lightly insert a rubber stopper while you are transferring the flask. You can also try to use a one-hole stopper temporarily plugged with a short section of a stirring rod.

8. Once (in your judgment) it reaches equilibrium, measure the temperature of the water in the bucket. This should take less than one minute.

9. The air has contracted and some water has entered the flask. Measure the volume of the water in the flask. (The most accurate reading occurs when the air pressure about the water is in equilibrium with the outside air. This can be established by raising the bottom of the flask so that the liquid level in the flask is at the same height as the liquid level in the bucket.) See Figure 91-2.

10. Subtract the volume of water from the total capacity of the flask.

11. Plot the two points you measured on a graph with volume on the y-axis and temperature on the x-axis. Leave enough room on both axes so that the point where the line connecting the points extrapolates to *zero* volume fits on the graph. Draw that line and determine the temperature where the volume would be zero.

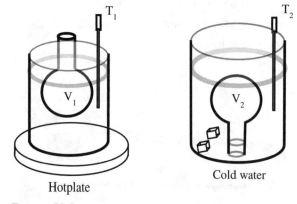

Hotplate

Cold water

Figure 91-2 *When the temperature cools from T_1 to T_2 the volume changes from V_1 to V_2.*

Expected Results

The accepted value for absolute zero is 0 K or –273°C or –459.7°F. However, because the first part of this experiment is a ballpark measurement, values anywhere from –175 to –350°C are reasonable extrapolations. Although this is a wide range, the concept that extrapolating the volume versus temperature curve until the volume goes to zero is significant. Statistically, we know we are on very shaky ground because we are generating data points that are far removed from the temperature we are extrapolating to.

Why It Works

Charles's law states that $T_1/V_1 = T_2/V_2$. This can be interpreted as saying that the volume of a gas is directly proportional to the temperature. The lowest temperature conceivable is the temperature when the gas contracts to a zero volume. This can never actually occur. If we extrapolate the linear relationship between temperature and volume to zero volume, you can determine a value for absolute zero.

Similarly, *Gay-Lussac's law* states that $T_1/P_1 = T_2/P_2$. Absolute zero is the temperature where the temperature pressure line is extrapolated to zero pressure.

Other Things to Try

The accuracy of this measurement can be improved on by using the Gay-Lussac's apparatus and extrapolating the temperature versus pressure curve to zero pressure, as shown in Figure 91-3.

Two different sets of measurements of temperature versus pressure are shown in Figure 91-4.

The point where the pressure extrapolates to zero is interpreted as absolute zero, as shown in Figure 91-5.

Figure 91-3 *The Gay-Lussac apparatus is used to measure the relationship between pressure and temperature.*

One source of error in determining the change in volume with temperature is the presense of water vapor in the flask. Use of oil instead of water to immerse the flask in avoids this problem. This is a messier but more accurate way to estimate absolute zero.

Absolute zero can be determined by measuring the speed of sound at different temperatures, as described in "Determining Absolute Zero Using a

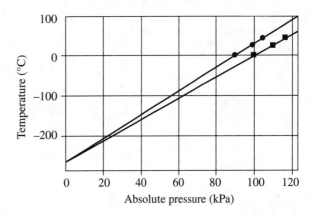

Figure 91-5 *Extrapolation of temperature versus pressure data to zero pressure to find absolute zero. Courtesy PASCO.*

Tuning Fork," by Jeffrey D. Goldader (*The Physics Teacher* 46, April 2008, 206–209).

The Point

Absolute zero cannot be measured directly. It can be determined by extrapolating measurements of pressure versus temperature to zero pressure. Similarly, absolute zero can be determined by extrapolating measurements of volume versus temperature to zero volume.

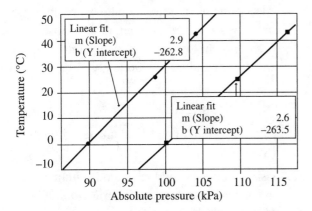

Figure 91-4 *Measurement of temperature versus pressure.*

Project 92
Liquid nitrogen.

The Idea

The air we breath is about 78% nitrogen. Typically this is a gas but when brought down to an extremely cold temperature nitrogen becomes a liquid. Not only is liquid nitrogen fun to play with, but it gives you an opportunity to begin to explore low-temperature physics.

What You Need

- safety goggles
- tongs and thermal mitt
- dewar (specifically designed to contain liquid nitrogen)
- sample of liquid nitrogen in a secure container
- plastic beaker
- table or other surface that will not be harmed by very cold temperatures
- Pyrex bowl large enough to hold a small quantity of liquid nitrogen and immerse the other objects listed here
- 12 inches (approximately) of lead tin or other solder wire
- 20 g hooked mass
- flowers
- banana
- hammer
- metal tube with one end open and the other sealed

- medicine bottle with easy snap on/snap off lid or a 35mm film canister
- cork that *loosely* fits into the open end of the metal tube
- balloon

Method

Safety

1. Liquid nitrogen *must* be handled safely. This means it *must* be stored in a specially designed container called a *dewar*, which is intended for this purpose. Do not put liquid nitrogen in a typical lunchbox thermos, which should *never* be used for liquid nitrogen.

2. Liquid nitrogen is so cold, it can freeze human skin in a very short time. For all these activities, all participants should wear safety goggles and avoid any sustained contact with skin. Be especially carefully to avoid splashing liquid nitrogen, which could get trapped under clothing and cause freezing. Also, remember, objects that have been immersed in liquid nitrogen have themselves been brought to a very low temperature and should be handled appropriately.

Splitting a banana

1. Carefully pour some liquid nitrogen into the Pyrex bowl.

2. Using the tongs and thermal mitt, immerse the banana into the liquid nitrogen for about 15–30 seconds. Initial bubbling and vaporization diminish as the banana

Figure 92-1 *Freezing a banana in liquid nitrogen. Courtesy PASCO.*

approaches equilibrium with the liquid nitrogen, as shown in Figure 92-1.

3. Remove the banana and place it on the table.

4. Take the hammer and strike the banana, as shown in Figures 92-2 and 92-3. (For contrast, you can strike an unchilled banana either before or after, showing the effect of the much-more-brittle frozen version.)

Frozen flowers

1. Using the same bowl as the previous experiment, immerse some fresh flowers in the liquid nitrogen.

2. Drop the bouquet on the floor or strike them on the table.

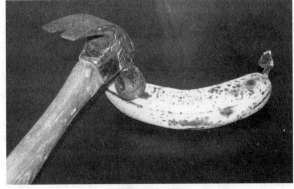

Figure 92-2 *Striking a deep-frozen banana with a hammer. Courtesy PASCO.*

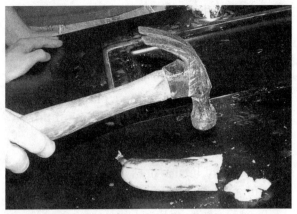

Figure 92-3 *Normally soft and pliable objects when deep frozen become brittle. Courtesy PASCO.*

Solder spring

1. Form the piece of solder into a coil roughly ¾ inch (1–2 centimeters) in diameter.

2. Notice the lack of stiffness in the spring.

3. Immerse the spring in the liquid nitrogen for roughly 15–30 seconds.

4. Hang a 20 g (or so) hooked mass from the frozen spring and compare its stiffness with that of the room-temperature version.

Balloon—filled with air

1. Blow up the balloon and tie a knot in the open end.

2. Immerse the balloon in the liquid nitrogen.

3. Observe what happens as the balloon is cooled.

4. Remove the balloon from the liquid nitrogen and again observe what happens.

Balloon—filled with liquid nitrogen

1. Pour a small amount (start with about 10 mL) of liquid nitrogen in a balloon.

2. Tie a knot in the open end.

3. Set the balloon on a table.

4. Step back and *make sure no one is near the (expanding) balloon* and especially *make sure no one's face is close to the balloon.*

5. Observe what happens as the balloon is exposed to the warmer air temperature.

Prescription container/Film canister

1. Place the film canister (or a plastic prescription container with a snap-off lid) on a table top or on the floor. *Do not* use a prescription container with a screw-on or a child-proof lid that does not easily snap off with moderate force.

2. Pour some of the liquid nitrogen into the plastic beaker.

3. Pour some of the liquid nitrogen from the plastic beaker into the film canister. Fill the film canister about *¼ full* with liquid nitrogen.

4. Snap on the top.

5. Stand back as pressure builds up in the container.

Cork gun

1. Place the cork gun where it is aiming in a *safe direction* (and specifically *not* directed toward anyone's face).

2. Pour about 50 mL of liquid nitrogen into the cylinder.

3. *Lightly* place the cork in the open end of the cylinder. Do not jam the cork in so tightly that it cannot be pushed out by the pressure that will build up in the cylinder.

4. Stand back. Pressure will build up as the liquid nitrogen evaporates.

Expected Results

The frozen banana and flowers will shatter. The solder will temporarily become much more spring-like. The air-filled balloon will shrink as the air inside contracts from the extreme cold, and then it will re-inflate as it warms up again. The liquid nitrogen-filled balloon will expand and possibly burst. The lids of the film canister/ prescription bottle will pop off. The cork will shoot out of the metal cylinder.

Why It Works

Objects become more brittle and contract from the extreme cold. As the liquid nitrogen evaporates, it occupies a much larger volume. For a given volume, the gas has a much larger pressure.

Other Things to Try

For many experimenters, liquid nitrogen may not be easily available on a daily basis. While you have a supply of liquid nitrogen available, you may want to consider doing the other projects that also require liquid nitrogen, such as Project 101 (effect of temperature on resistance) and Project 106 (superconductivity).

The Point

Liquid nitrogen provides an opportunity to explore low-temperature physics. This includes making normally elastic materials brittle. Materials cooled by liquid nitrogen contract. As liquid nitrogen evaporates, it expands.

Project 93
Boiling water in a paper cup.

The Idea

Is it possible to boil water over a flame in a paper cup? This project lets you find out why this is possible.

What You Need

- 2 paper cups—most "paper" cups have a thin coating of wax, which can still be used, but if you can get them, uncoated cups are preferable
- water
- flame—a match or a Bunsen burner
- thermometer or digital temperature sensor
- Styrofoam cup
- sand (enough to partially fill a paper cup)
- water balloon
- paper bag

Method

1. Fill the paper cup nearly to the top with water.
2. Hold the cup over the flame.
3. Continue doing this until either the paper burns or the water boils. Optional: measure the temperature as it is heating up.
4. Fill the second paper cup with sand.
5. Hold this cup over the flame and observe the effect of the flame on the cup.

6. Fill the Styrofoam cup nearly to the top with water.
7. Hold this cup of the flame and observe the effect of the flame on the cup, as shown in Figure 93-1.

Expected Results

The water will boil in the paper cup. If the cup is coated with wax, the wax may melt, especially above the waterline. If there is a circular rim on the bottom, it may burn without burning through the cup. The paper cup filled with sand will char, but it won't necessarily burst into flames. The Styrofoam will melt and, where the flame is applied, possibly leave a hole in the side of the cup.

Figure 93-1 *Boiling water in a paper cup.*

Why It Works

When heat is added to water, its temperature increases until it reaches the boiling point of water at 100°C. The paper doesn't burn because heat is conducted away from the paper before it can reach its kindling point (the temperature where it begins to burn). Paper begins to burn at around 233°C (which is close to the nominal value 451°F for paper, made famous in Ray Bradbury's novel *Fahrenheit 451*). The water temperature can increase until it boils and still remain well below the kindling temperature of paper.

Sand conducts heat away from the paper. However, unlike the paper, sand does not undergo a phase changes as water does at its boiling point. The temperature increases above 100°C. This is why we see charring in the paper cup containing sand.

The Styrofoam is an insulator. As a result, the water does not conduct heat away from the Styrofoam cup as it does with the paper cup, which conducts heat much more readily. This explains why the flame melts the Styrofoam.

Other Things to Try

An alternative approach is to wrap a piece of paper around a metal pipe and note its response to a flame. In a similar manner, the metal pipe conducts heat away from the paper before it can start to burn.

Fill the paper bag with water and hold it over the flame. The water conducts heat away from the paper at a fast enough rate to keep it from burning.

The Point

Phase changes in matter, such as the transition from liquid to vapor, take place at a constant temperature called the *boiling point*. A liquid cannot exceed the boiling point until all the liquid has evaporated. Materials such as sand conduct heat much better than air. Some materials such as Styrofoam are much better insulators than other materials, such as paper.

Project 94
Boiling water with ice.

The Idea

In this project, you use a piece of ice to cause a container of very warm water to start boiling. This is definitely not what most people would expect.

What You Need

- Pyrex Erlenmeyer flask (or a Florence flask with a partially flat bottom)
- rubber stopper (without holes)
- beaker tongs (or oven mitt)
- water
- few ice cubes
- hotplate
- ring stand with a ring small enough to support the flask upside down
- optional: bell jar and vacuum pump, beaker

Method

1. Partially fill the flask with water. There should be a gap of an inch or two above the water level when it is upside down.

2. Place the flask on the hotplate, as shown in Figure 94-1.

3. Keep the flask on the hotplate until the water boils.

4. Remove the flask from the hotplate.

5. Without delay, put the rubber stopper (snuggly) in the flask, *carefully* turn it upside

down, and place it in the ring. Use an oven mitt or tongs to handle the flask.

6. The water (having cooled slightly) should now be still quite hot, but no longer boiling.

7. Position a few ice cubes on the flat of the flask and observe.

Expected Results

Shortly after the ice cubes are placed on the bottom of the flask, bubbles start to emerge from

Figure 94-1 *Bringing a flask filled with water to (just under) boling.*

245

the top (near the stopper). These bubbles continue and the water in the flask boils for a short time. Careful observation should convince anyone watching that the bubbles are coming from the liquid itself and are not a leak in the rubber stopper. See Figure 94-2.

Why It Works

When a vapor (such as the air/water vapor mixture) is cooled, it contracts. As the volume of gas above the hot water decreases, the pressure also decreases. Water boils at 100°C (212°F) at standard atmospheric pressure, but at a slightly *lower* temperature when the pressure above the liquid is reduced.

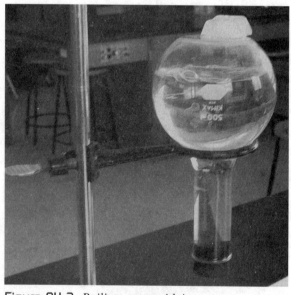

Figure 94-2 *Boiling water with ice.*

Other Things to Try

You can try this another way:

1. Fill a beaker with water.
2. Place it on a hotplate and bring it to a boil.
3. Remove the beaker from the hotplate and let it cool until the boiling just stops.
4. Place the beaker in a vacuum chamber (bell jar on a vacuum plate).
5. Attach and turn on the vacuum pump to evacuate the chamber.
6. Compare the effect of directly applying a vacuum to the reduced pressure caused by the ice.

The Point

Water boils at a lower temperature when the pressure of the air above it is lowered.

Project 95
Seebeck effect/Peltier effect. Semiconductor heating.

The Idea

Much of physics concerns itself with how one form of energy is changed into another. This experiment explores how heat can cause an electrical current to flow. Although this is not yet efficient enough to be used as a significant source of electrical power, it is widely used in the form of thermocouples to measure temperature. This is known as the *Seebeck effect*.

The reverse—where electrical current flowing through certain materials results in one part of the circuit getting hot and the other part getting cold—is known as the *Peltier effect*. Unwanted heat is dissipated by electronic components. These components must be cooled to function correctly. Peltier coolers have no moving parts and have been used to cool high-speed computer microprocessors. They are also used instead of dry ice in cloud chambers. (See Project 125.)

What You Need

- voltmeter (or multimeter configured as an voltmeter)
- ammeter (or multimeter configured as an ammeter)
- variable DC power supply
- jumpers with alligator clips
- 1000 ohm resistor

- 2 4-inch lengths of various types of (uninsulated) metal wire, including iron, copper, constantan, and aluminum
- heat source, such as a candle, Bunsen burner, or a soldering iron
- ice cubes
- optional: 2 thermocouples to be used as temperature sensors

Method

Seebeck effect

1. Select two different wire materials. Take two pieces of the first material and one piece of the second material. Let's say we start with two pieces of iron and one piece of copper.

2. Attach each end of the copper wire to each of the two pieces of iron wire by twisting about a one-half inch length of the wire together.

3. Connect the two unattached ends of the iron wire to the positive and negative terminals of the voltmeter. See Figure 95-1. Set the voltmeter on the most sensitive setting. The 250 mV (0.250 V) range is a good place to start.

4. Measure the voltage at room temperature. (Momentarily disconnect one of the voltmeter connections to verify that the voltage you are reading is the result of the circuit you set up, rather than a small stray voltage reading.)

5. Touch one junction (twisted wire connection) to the ice, leaving the second junction at room temperature. How does that affect the voltage reading?

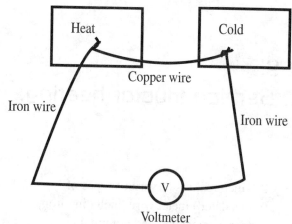

Figure 95-1 *Wiring arrangement for measuring the Seebeck effect.*

6. Place the second junction in the heat source and see what happens. Be careful because the various metal wires conduct heat and can burn anything that comes in contact with it.

7. Try this with as many material combinations as you can.

8. Place a temperature sensor on each of the twisted metal junctions. Vary the junction temperatures. Plot the voltage as a function of the *difference between* the two junction temperatures.

Peltier effect

1. As before, connect each end of one piece of wire to a piece of a second type of metal wire.

2. Connect the components as a series circuit consisting of the wires, the ammeter, a 1000 ohm resistor, and an ammeter. This circuit is shown in Figure 95-2.

3. Adjust the DC power supply, so about 10mA (10 milliamps or 0.01 amps) is flowing through the circuit. (You can use a 9-volt battery instead of an adjustable DC power supply, which results in slightly less than this current with the 1000 ohm resistor.)

4. Put water on each of the junctions. What happens? Reverse the direction of the current flow by exchanging the wire connected to the

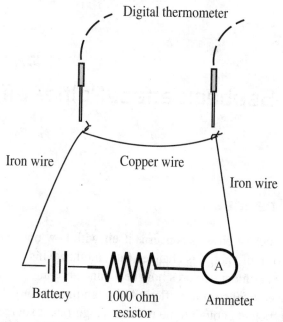

Figure 95-2 *Apparatus for the Peltier effect.*

DC power supply. How does that affect what you find?

5. Monitor the temperature of each of the junctions as you vary the current flowing through the circuit. Plot the temperatures of each junction and the difference versus the current.

Expected Results

In the Seebeck effect, a temperature difference at the two junctions results in a voltage generated through the circuit.

The Peltier effect results in temperature differences at the junctions when a current flows through the circuit.

Why It Works

A temperature difference between two dissimilar metals results in an electrical potential that drives a current through a circuit. The reverse effect causes a temperature difference when a current flows.

Other Things to Try

Commercial thermocouples that employ dissimilar metals are based on the Peltier effect and can be used to study this principle. Some temperature control devices that serve as a means of studying the Seebeck effect are available commercially.

The Point

The Seebeck effect and Peltier effect describe a set of interactions between the thermal properties and the electrical properties of matter.

Section 9

Electricity and Magnetism

Project 96
Static charges.

The Idea

According to Newton's law of universal gravitation, any mass exerts a force on any other mass. Electric charges work in a very similar way. The farther away you get, the weaker the force. Because the electric force is so much stronger than the gravitational force, it is much easier to measure. This experiment explores the nature of the electrostatic force and establishes the basis for Coulomb's law.

What You Need

- 2 pith balls or conductively coated Styrofoam balls (conductively coated ping-pong balls are also an option)
- 2 pieces of string about 16 inches in length
- movable ring stand with a pendulum clamp (or other horizontal support)
- small *nonconductive* post on a stand (the post should be a few inches in length and consist of a thin wooden dowel or a short glass or plastic rod)
- ruler

- rubber rod/wool pair (or equivalent) to apply a charge to the pith balls
- optional: light source to project the image of the pith balls onto a screen (an overhead project or LCD projector can serve this purpose)

Method

1. Attach one side of each of the two strings to the pith ball.
2. Attach the other sides of the string to the pendulum clamp separated by a few inches, so the pith ball can swing in only one direction, as shown in Figure 96-1.
3. Attach the other pith ball to the nonconductive stand.
4. The swinging pith ball should be positioned so it can only swing closer to and further from the stationary ball.
5. Draw a reference mark on the bottom of the ring stand to indicate the rest position of the swinging pith ball without being subjected to any force other than gravity.

6. Vigorously rub the wool against the rubber rod to charge it up. Touch each of the pith balls to apply the same charge to them. Touching the two balls together will make the charges nearly equal, but it is not necessary to do this.

7. Start with a distance between the pith balls that allows the swinging ball to hang vertically.

8. Slowly bring the swinging pith ball closer until the repulsion between the two pith balls causes the swinging pith ball to move away from the stationary ball.

9. Measure how far the swinging pith ball moves *horizontally*. You may do this by observing from above and measuring the distance the ball has moved from the reference point.

10. Record the horizontal distance between the centers of each of the pith balls.

11. Repeat this measurement a few times by moving the swinging pith ball in a little closer.

12. The horizontal separation, *x*, between the unconstrained pith ball and its equilibrium position is a good indication of the force. (This can actually be worked out in terms of the force, but this is unnecessary to explore the key point of this experiment.) For small angles that the pith ball makes with the vertical, the electrostatic force is directly proportional to the separation from equilibrium.

13. The separation between the stationary ball and the equilibrium positions is designated as *d*, as shown in Figure 96-2. The total distance between the two pith balls is given by d + x. Make a graph of the separation from equilibrium, *x*, and the distance, d + x, between the balls.

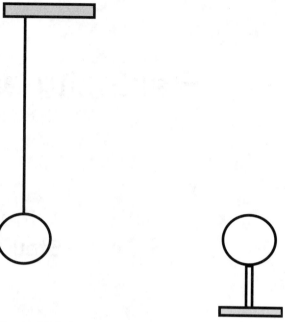

Figure 96-1 *Two uncharged balls in equilibrium position.*

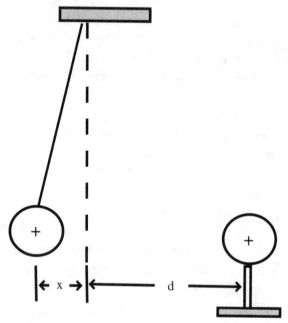

Figure 96-2 *Two charged balls show displacement from equilibrium position.*

Expected Results

The closer the two balls get, the greater the force.

This relationship is not linear.

The closer the balls get, the faster the increase in force between the balls.

Specifically, this is an inverse square relationship.

Overall, this experiment works best on a day with low humidity.

Why It Works

Coulomb's law states that the force (in newtons) between two charges, q_1 and q_2 (in Coulombs or C), separated by a distance, d (in meters), is given by:

$$F = \frac{kq_1q_2}{r^2}$$

where k is the Coulomb constant = 9.0×10^9 m²/C².

Other Things to Try

Finding how many electrons are on a charged balloon.

Hang two balloons after first determining their mass. Charge them and touch them together, so they have roughly the same charge. The Coulomb force results in the balloons repelling and separating, as shown in Figure 96-3.

The number, n, of electrons on each of the balloons can then be determined from:

$$n = \frac{1}{q} \sqrt{\frac{mg \tan\theta}{k}}$$

where q is the charge on 1 electron = 1.6×10^{-19} C

k is Coulomb's constant = 9.0×10^9 m²/C²

m is the mass of each balloon

and θ is the angle the string of each balloon makes with a vertical line.

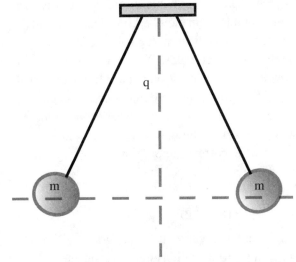

Figure 96-3 *Two balloons with like charges repel.*

The Coulomb force can also be explored in the following simple demonstrations:

Chasing a can

A charged rubber rod, such as the one previously used to charge the pith balls, induces a current in a can. This enables you to roll the can back and forth across the table, as if the rod were a magic wand.

Bending water

A thin stream of water from a facet can be bent by a charged rod.

Tape

The simplest of all is to take some transparent tape and adhere it to a table. After pulling it up, the tape will have acquired a charge that will be attracted or repelled by other nearby objects. Two similar pieces of tape acquire similar charges and are, therefore, repelled.

Plasma globe/Sunder ball

This is a true "evil genius" prop that looks like something out of a Frankenstein movie. A small *tesla coil* produces a large voltage difference inside a glass bulb. This is similar to the way that charge builds up in clouds. Mini lightning bolts discharge through an inert gas in the bulb, producing an eerie glow. The electrons in the gas flow to the electrical ground harmlessly provided by a finger touching the outer edge of the glass as if the person touching the outside of the globe was a human lightning rod. This is safe to do because the current (amps) flowing is very small. See Figure 96-4.

Further Analysis

Apply an Excel curve fit for the graph of the separation from equilibrium, x, versus the separation $(d + x)$ between the balls. A scatter plot with the power option selected should indicate the best fit closest to –2, which is an inverse square relationship.

You could also plot x versus $1/(d + x)^2$. A linear fit to this graph would indicate an inverse square relationship between x (which is an indicator of the magnitude of the force) and d.

Figure 96-4 *Plasma globe.*

The Point

The force between two charges is directly proportional to the product of the charges and inversely proportional to the distance separating the charges as given by Coulomb's law.

Project 97
Making lightning. The van de Graaff generator.

The Idea

A *van de Graaff generator* builds up static electric charges, which when discharged, produce visible lightning-like sparks. This is one of the more memorable physics experiments and provides a great introduction to the basic ideas of electricity.

What You Need

- van de Graaff generator
- grounding sphere
- grounding wire—with alligator clip terminations or just plain insulated wire with the ends stripped
- confetti, paper holes from a hole puncher, Rice Krispies
- strips of paper
- scotch tape
- a few aluminum pie pans (the small ones that are about 4 inches in diameter are preferable)
- electroscope
- glass rod and silk, rubber rod, and wool or fur
- short (12 inch or so) neon light bulb
- insulated (plastic) crate
- other van de Graaff toys: spinner, spark gap
- electroscope

Method

A word about safety

When used as intended and according to manufacturer's specifications, this is a safe experiment. The van de Graaff generator produces high voltages. These can be over 20,000 volts, which may sound high. However, similar voltages are produced by scraping your shoes across a carpet on a day of low relative humidity. The voltage is high, but the current is very low, so the high voltage is not hazardous because very few electrons are involved.

Remove all electrical devices from your pockets or wrist. Set up a safe area with unobstructed access to the van de Graaff generator.

Here are some considerations:

- Sparks can damage electronic devices, including cell phones, audio devices, calculators, computers, digital watches, and pacemakers.
- This activity is fun because of the dramatic and sudden static electric discharges. However, make sure no one is put at risk by anyone's sudden reactions to this apparatus.
- Be careful not to build up higher-than-intended charges using long human chains or other storage devices, such as capacitors or Leyden jars.
- Make sure sparks are not near flammable materials, such as natural gas pipes or combustible laboratory chemicals.

Lightning

1. Place the van de Graaff generator on a table and plug in the electrical chord.

2. Position a grounding electrode a few centimeters from the conducting sphere of the van de Graaff generator. If you don't have a discharge sphere purchased for this purpose, you can improvise using a metal rod, such as a ring stand.

3. Attach a wire between the grounding electrode and an electrical ground, such as a water pipe or a metal beam that is part of the building structure.

4. Turn on the van de Graaff generator.

5. Darken the room.

6. Move the conducting sphere of the van de Graaff generator back and forth, and observe what is the maximum distance a discharge will cross.

7. Place a sheet of paper in the path of the spark. Does a sheet of paper stop the spark?

8. Turn off the generator and touch the grounding sphere to the conducting sphere of the van de Graaff generator to remove any residual charge.

Paper strips

1. Ground the conducting sphere.

2. Use tape to attach strips of paper to the top of the conducting sphere of the van de Graaff generator.

3. Turn on the generator and observe the paper hairs separating and standing on end, as shown in Figure 97-1.

4. Turn the generator off and ground the conducting sphere with the ground electrode to remove any residual charge.

A hair-raising experience

1. Ground the conducting sphere.

2. Place an insulating surface on the floor near the van de Graaff generator. An inverted

Figure 97-1 *Like charges repel, causing the pieces of paper to separate.*

plastic crate works well for this purpose. The reason for doing this is to make sure no discharge occurs to conductors, such as water pipes, under the floor.

3. Place your hand on top of the generator with the palm of your hand face down.

4. Have someone turn on the generator. You can do it yourself, but be prepared for the possibility of a mild and harmless shock.

5. This works best with people with long, fine hair on low humidity days. If someone comes close, they could be mildly shocked.

6. When you are finished, turn off the van de Graaff generator, step off the crate, and discharge the conducting sphere.

Levitating pie pan

1. Ground the conducting sphere.

2. Place an aluminum pie tin on the conducting sphere.

3. Turn on the generator and observe the result.

4. Try this with several pie pans stacked on top of the conducting sphere.

5. Turn the generator off and ground the conducting sphere with the ground electrode to remove any residual charge.

Positive or negative?

1. Ground the conducting sphere.

2. Turn on the generator and bring the electroscope near the conducting sphere. Don't be totally surprised if you get a spark.

3. Rub the glass rod with the silk to produce a negative charge on the glass rod.

4. Bring the negatively charged glass rod to the top of the electroscope to separate the leaves.

5. Bring the negatively charged electroscope near the van de Graaff generator and observe whether the leaves separate further or move closer together.

6. Rub the rubber rod with the wool (or fur) to produce a positive charge on the glass rod.

7. Bring the positively charged rubber rod to the top of the electroscope to separate the leaves.

8. Bring the positively charged electroscope near the van de Graaff generator and observe whether the leaves separate further or move closer together.

9. Based on the response of the electroscope, what do you conclude about the type of charge produced on the conducting sphere of the van de Graaff generator?

Neon bulb

1. Turn on the van de Graaff generator.

2. Darken the room.

3. Bring a small neon bulb (with no electrical connections to either end) near the conducting sphere. If you get too close, some sparks will likely discharge harmlessly in your hand. If you prefer for this not to happen, you can rig up a nonconducting holder to support the neon bulb during this exercise. Be careful to avoid sudden moves that might result in dropping the bulb.

4. Position the neon bulb parallel to the floor on a line pointing toward the center of the conducting sphere. Move it closer and further from the van de Graaff and observe.

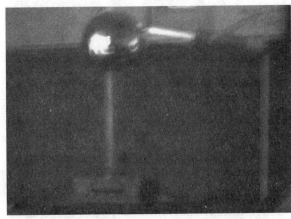

Figure 97-2 *Lighting a neon bulb by exposing it to an electric field.*

5. Position the neon bulb parallel to the floor on a line pointing perpendicular to the circumference of the conducting sphere, as shown in Figure 97-2. Move it closer and further from the van de Graaff and observe.

6. Turn the generator off and ground the conducting sphere with the ground electrode to remove any residual charge.

Expected Results

Sparks will discharge through the air. Longer sparks are generated if the voltage is high and the humidity is low. Best-case spark length is shown in Figure 97-3.

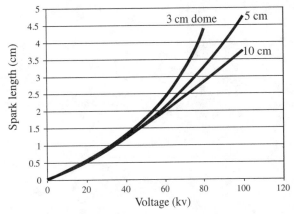

Figure 97-3 *Spark length. Based on data from A. D. Moore,* Electrostatics: Exploring, Controlling, and Using Static Electricity *(New York: Doubleday-Anchor, 1968).*

Why It Works

An electrical motor drives a nonconducting belt that separates positive and negative charges. Negative charges are drawn away from the conducting metal sphere at the top of the device, leaving a static positive charge on the sphere. Bringing a neutral or negatively charged object close to the conducting sphere results in a discharge in the form of a lightning-like spark through the air. The maximum length of the spark depends on the relative humidity in the air. As a rule of thumb, the spark jumps about 1 centimeter for every 10,000 volts that build up.

Objects that come into contact with the conducting sphere, themselves become charged. This causes repulsion between pieces of paper taped to the conducting sphere, of the hair of a person touching the sphere (while standing on an insulating surface), and between small objects, such as Styrofoam chips or Rice Krispies that come into contact with the sphere.

Other Things to Try

Charges separated by the van de Graaff generator can be concentrated on the bulb of an electroscope. The sign of the charge can be determined by observing the effect of a charged rod brought near the bulb. If the charge is the same, the leaves of the electroscope are driven further apart. If the charge is different, the leaves are drawn closer together. A positively charged rod results by rubbing wool or fur on rubber. A negatively charged rod results by rubbing silk on glass.

Use of a *tesla coil* is another way to generate sparks that jump through the air. These can be purchased as hand-held units.

The Point

This experiment shows how, through the movement of dissimilar materials against each other, static electric charges can build up. An electric field is established in the region separating the electrical charges which can force the electrons to move.

Project 98
The Wimshurst machine. Separating and storing charges.

The Idea

The *Wimshurst machine*, like the van de Graaff generator, is capable of throwing long sparks as much as several centimeters between two small conductive spheres. The Wimshurst machine is usually also tied to a Leyden jar, which presents a good opportunity to explore capacitance and charge storage.

What You Need

- Wimshurst machine

Method

1. CAUTION: Electrical circuitry including pacemakers, hearing aids, cell phones, flash drives, electronic car door locks, and computers may be damaged by the sparks generated in this experiment. *In addition, follow all safety instructions provided by the manufacturer of this device.*

2. Set the Wimshurst apparatus on a table.

3. Darken the room.

4. Make sure all electrical jumpers are in place for your particular setup. Check with the manufacturer's instructions to make sure the apparatus is set up properly with correct electrical paths to the Leyden jars and discharge spheres. You can do this both with the Leyden jars connected or not connected to your circuit.

5. Separate the two discharge spheres by more than 8 cm.

6. Turn the handle for five seconds or so, as shown in Figure 98-1.

7. Holding only the insulated wooden handles for the discharge spheres, slowly bring them together.

8. Note the distance between the discharge spheres when the first discharge occurs.

9. Set the discharge spheres at roughly the same slightly closer and slightly further distances.

10. If the Leyden jars were in your circuit, repeat to see what happens when they are not connected.

11. Touch the two spheres together for a few seconds when finished to make sure no residual charges are on the electrodes.

Figure 98-1 *Chris Aleo demonstrates the operation of a Wimshurst machine.*

Expected Results

With the discharged spheres separated by a large distance, nothing should happen.

As the spheres approach to within a few centimeters separating them, a lightning bolt will jump across the gap, as shown in Figure 98-2.

Why It Works

The Wimshurst machine is constructed from two parallel plates made from insulating material such as Lucite or glass. The plates are arranged to be turned by hand in opposite directions. Narrow metal strips are mounted on the plates and oriented along the radius. Charges are transferred by metal brushes that sweep across the metal strips as the plates rotate. In contrast to the van de Graaff generator, the Wimshurst machine separates charge by the principle of induction rather than friction. Positive and negative charges accumulate, and they can either charge a Leyden jar or discharge across a gap.

Other Things to Try

Charges separated by the Wimshurst machine can be determined using an electroscope, as described in the previous experiment.

The Point

This experiment shows how, through the movement of two insulating plates near each other, static electric charges can build up. The separated charges can be stored or discharged across a small nonconductive gap.

Figure 98-2 *Wimshurst machine electrical discharge.*

Project 99
Running into resistance. Ohm's law.

The Idea

Ohm's law forms the basis for understanding how electricity flows through circuits. This is a very simple relationship that involves three things: 1) the *voltage* or the push that move electrons through the circuit, 2) the *current* (or amps), which is a measure of how much electricity is flowing through that circuit as a result of that push, and 3) the *resistance* (in ohms), which does all it can to make it difficult for the electricity to flow.

Because of its simplicity, this experiment is a good one for you to discover the law for yourself, based on your measurements.

What You Need

- one 100 ohm resistor rated for 0.5 watt (other resistor values can work, but the resistor must be rated to handle the wattage that will be applied to it; the wattage is supplied by the resistor manufacturer and is often marked on the resistor)
- ammeter
- voltmeter
- DC power supply (or battery)
- wires to connect to battery terminals

Method

1. Connect a circuit, as shown in Figure 99-1. This is a circuit consisting of a resistor in series with a DC power supply and an ammeter with a voltmeter connected by jumper wires to each of the ends of the resistor. A drawing called an *electrical schematic* is shown in Figure 99-2. This is equivalent to Figure 99-1, but it shows the electrical connections without regard to the actual physical layout of the components.

2. Turn the DC power supply to zero.

3. Set the ammeter to read milliamps. Set the voltmeter to read 0–10 volts.

4. Increase the DC power supply to give a voltage reading of 0.2 volt.

5. Read the current.

6. Do the same with a voltage of 0.4, 0.6, and 0.8 volts.

7. Graph current versus voltage. Draw a line that best fits the data. What is the significance of the slope of the line?

8. Repeat this with a 200 and a 300 ohm resistor.

Expected Results

For a given resistor, the greater the voltage, the more current flows.

As resistance increases, less current flows for a given voltage.

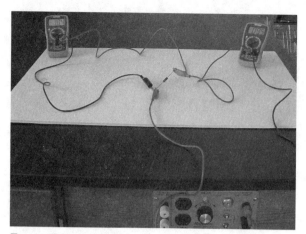

Figure 99-1 *Circuit for measuring Ohm's law.*

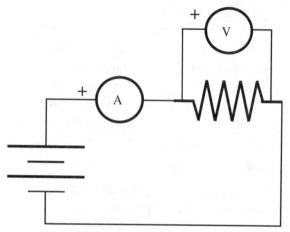

Figure 99-2 *Schematic for measuring Ohm's law.*

Voltage increases linearly with current. The slope of the line is the resistance the current is flowing through.

Why It Works

Ohm's law is given by volts = resistance (ohms) × current (amps).

From this, you can see that the slope of the volts versus the current graph is resistance.

Other Things to Try

What happens if you have two or three resistors of the same resistance in a row, one connected to the next? This is called a *series circuit* and is shown in Figure 99-3. For a given voltage, is the current greater or less than for a single resistor?

What happens if you take those same three resistors and connect them in a parallel circuit, as shown in Figure 99-4?

The Point

Ohm's law relates the voltage, current, and resistance of a circuit. The voltage at any particular time equals the current times the resistance.

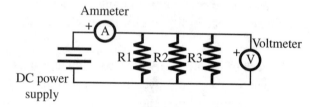

Figure 99-3 *Measuring resistors in a series circuit.*

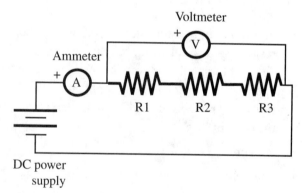

Figure 99-4 *Measuring resistors in a parallel circuit.*

Project 100
Circuits: Bulbs and buzzers.

The Idea

If you have never built a circuit before with your own hands, this is your chance. Like many of the experiments in this book, various levels of complexity exist and you can take the experiment as far as you care to. You start with building a simple circuit, such as making a bell ring. Then, you build a basic telegraph system. You branch out and add series and parallel paths to simple circuits. Next, you measure the current and voltage at various points in the circuit. Finally, you look at how Ohm's law can be applied to more complicated circuits.

What You Need

- jumper wires
- 6 Christmas tree bulbs (or low-voltage bulbs and sockets)
- various (low-voltage) electrical devices such as bells, buzzers, LEDs
- DC power supply (or battery as in the previous project)
- knife switch
- ammeter (or multimeter set up as an ammeter)
- voltmeter (or multimeter set up as an voltmeter)
- for the telegraph: 5–10 feet of insulated wire, iron nail, two blocks of wood roughly 3 × 6 × ¾ inches, a second block of wood ¼ inch taller than the nail after being nailed into the block, a "tin" can, tin snips, and a few small nails

Method

Building a circuit

1. Look at the circuit diagram in Figure 100-1 and make the appropriate connections.

2. If you have an adjustable DC power supply, set a voltage of 2–3 volts and keep it constant throughout the test. You may need to adjust this, depending on the circuit you are working with.

3. The circuit diagram should give you all the information you need. Here are a few details that may be helpful:

   - Attach a jumper wire to the positive and negative terminals of the DC power supply or battery. (Not that the electrons care, but red is generally used for positive and black is used for negative for clarity in assembling the circuits.)

   - There must be a complete path from the positive of the power supply and back to the negative.

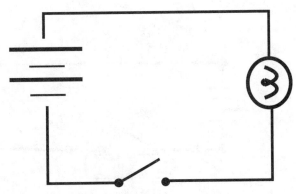

Figure 100-1 *Basic electrical circuit consisting of a battery, bulb and a switch.*

– All connections must be metal-to-metal. If insulation is on the wires, you either need to use a bare-metal alligator connector or remove the insulation.

Making a telegraph

1. Wind 25–50 turns of insulated wire around a large iron nail. Leave the two ends of the wire free and remove about ¾ inch of insulation.

2. Hammer the nail into the wooden block.

3. Cut two strips from your can, roughly 2½ inches long × ½ inch wide. Both pieces should be flexible.

4. Attach one of the metal strips to the wooden block. The height of the wooden block should be ¼ inch higher than the nail.

5. Attach the block with the metal strip to the base block, so the metal is above the head of the nail, but not touching.

6. Build the "key" by nailing the second metal strip to the second block on one side, and then putting a nail underneath the other end of the metal strip. Leave enough of the nail head exposed above the wood surface, so you can wrap wire around it.

7. OK. Let's hook everything up. What you have is a series circuit from the DC power supply through the electromagnet to the key. The key is simply a switch. When it closes, the electromagnet pulls the metal strip down, as shown in Figure 100-2. Short and long durations are the dots and dashes of Morse code. If you have enough wire, you separate the key and the receiver by some distance. (Note: if you use this to cheat on tests in school, please make sure you don't say you got the idea to do it here.)

Series and parallel circuits

1. Attach each of the two ends of a Christmas tree bulb (or a low-voltage bulb in a socket) across the power supply. (This means one of the wires attached to the bulb goes to the positive terminal and the other end goes to the negative lead.)

2. Connect three (or more) bulbs in series. Compare the brightness of these bulbs with the brightness of a single bulb.

3. Connect three bulbs in parallel and compare with the brightness of bulbs in a series and a single bulb.

Measuring the circuit

1. Repeat the previous set of measurements, but this time, include an ammeter in series with the circuit and a voltmeter in parallel with the circuit, as shown in Figure 100-2. It helps with the comparison if you keep the voltage constant throughout these measurements and compare the current flowing in the circuits.

2. Compare the current flowing in each the situations.

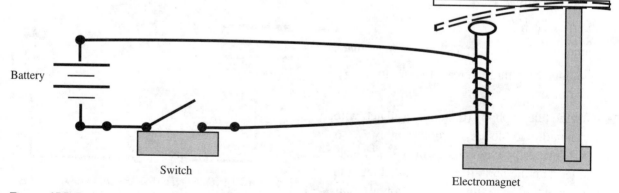

Battery

Switch

Electromagnet

Figure 100-2 *In a telegraph an electromagnet is activated when a switch is closed to complete the circuit.*

3. Apply Ohm's law (in the form $R = V/I$) to find the resistance of each of the circuits you measured.

Expected Results

Current will flow in a circuit if a continuous path exists from the positive terminal of the power supply through all components of the circuit, and then back to the negative terminal of the power supply.

Components in series *reduce* the current that can flow by effectively adding resistance to the circuit.

Components in parallel result in *increased* current flowing through the circuit. The resistance of the overall circuit is reduced when components are added in parallel.

Why It Works

When components are added in series, the voltage is distributed over all the components. As a result, less current is able to flow.

When components are added in parallel, alternate paths are provided for the current to flow back to the battery. For a given voltage, the push from the battery is able to force more current through the larger number of paths.

Other Things to Try

A next logical step is to create and test more complex networks of resistors. The following shows some examples. These can be analyzed using the following principles:

- Resistors in series simply add: $R_{series} = R_1 + R_2$
- Resistors in parallel add in a more complex fashion. Resistors in parallel can be thought of as one single equivalent resistance given by: $1/R_{parallel} = 1/R_1 + 1/R_2 + \ldots$
- These circuits can then be simplified by combining series and parallel circuits, and then applying Ohm's law in its various forms ($V = RI$, $R = VI$, and $I = V/R$).

The Point

Ohm's law determines how much current (or amps) flows through a circuit. For a given resistance (ohms), the greater the voltage, the greater the current.

Project 101
How does heat affect resistance?

The Idea

When matter heats up, the atoms start moving faster, like cars in rush-hour traffic. The hotter it gets, the more difficult it is for electrons to make it through a wire. In this experiment, you explore what happens when a conductor gets hot.

What You Need

- one 10-ohm resistor rated for 1 W or greater
- DC power supply or battery
- (optional) digital temperature sensor (some multimeters come with a thermocouple and setting to read the temperature)
- stopwatch

Method

1. Set up the circuit, as shown in Figure 101-1.

2. Attach the temperature sensor to the resistor. (Not to worry. If you don't have a thermocouple, there is still a way for you to do this.)

3. Set a voltage of about 5 V.

4. Measure the voltage, current, and temperature.

5. Continue taking readings of voltage, current, and temperature at regular intervals. If you don't have a temperature sensor, you can still proceed, taking note of the fact that the resistor is heating up qualitatively. A relationship between resistance and time (rather than temperature) can still be established.

6. Use Ohm's law to determine the resistance by dividing the voltage by the current (in amps).

7. What happens to resistance as the resistor heats up?

Expected Results

The higher the temperature, the higher the resistance. Resistance increases linearly with temperature. See Figure 101-2, which shows the resistance of a 5 centimeter (0.05 m) section of 20AWG copper wire over a range of temperatures.

Why It Works

When a conductor is heated, the molecules move in place more rapidly. Like a car moving on a highway with increasing traffic, the electrons cannot move as freely through the conductor. The result is the resistance increases.

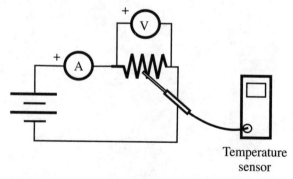

Temperature sensor

Figure 101-1 *Circuit for measuring the effect of heat of resistance.*

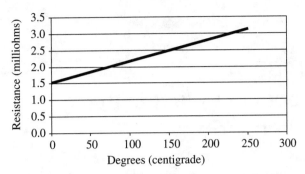

Figure 101-2 *Resistance versus temperature for 5 cm length of 20 AWG copper.*

Other Things to Try

Use an external source of heat, such as a hotplate or a Bunsen burner, to heat the (uninsulated) wire.

Use ice, dry ice, or liquid nitrogen to produce low temperatures. Your thermocouple may not read over the entire temperature range of your sample, but you can still obtain some extremely low-temperature readings as the sample warms.

The Point

Electrical resistance increases with temperature. This relationship is linear over a broad range of temperatures.

Project 102
Resistivity. Can iron conduct electricity better than copper?

The Idea

Yes, if the wire is longer or thicker. Copper is well known as a good conductor of electricity. This same is not usually said about iron. This project deals with two ideas that sound similar, but that are quite different: resistance and resistivity.

What You Need

- uninsulated copper wire 25 cm in length
- uninsulated iron wire 25 cm in length of the same diameter (this can be indicated by the wire gauge or AWG)
- (other material combinations, such as aluminum or silver wire can be used instead of, or in addition to, copper and iron)
- DC power supply
- ammeter
- voltmeter
- (if you have a digital multimeter, you may be able to use the ohmmeter setting directly)
- connecting wire
- ruler

Method

1. Set up the circuit as shown in Figure 102-1. Mark the wire with a Sharpie in 2 cm (or other convenient) lengths.

2. The ammeter is attached across the entire length of the wire. The current from the power supply flows through the entire length of the wire. The voltmeter is attached only across the selected length (2 cm, 4 cm and so on).

3. Read voltage, current, and distance.

4. Find the electrical resistance from Ohm's law by dividing the voltage (volts) by the current (amps). This gives a resistance reading in ohms. This can also be directly read from an ohmmeter if you have one.

5. Compare the resistance you measure for different lengths.

6. For a given diameter, multiplying the resistance by the length gives a measure of the wire's resistivity. What do you find happens to this value as the length increases?

Expected Results

The longer the wire, the greater the resistance.

The greater the cross-sectional area of the wire, the lower the resistance.

Resistance increases (linearly) with length.

Resistance is inversely proportional to cross-sectional area. This is represented in Figure 102-2.

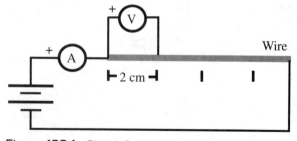

Figure 102-1 *Circuit for measuring resistivity.*

Section 9: Electricity and Magnetism

268

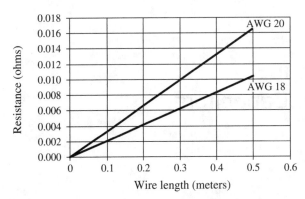

Figure 102-2 *Resistance of various lengths of copper wire.*

The cross-sectional areas of the various American wire gauges (AWG) are shown in Table 102-1.

Table 102-1

AWG	Cross-sectional area (mm²)
22	0.644
20	0.812
18	1.024
16	1.291
14	1.628

Resistivity at 20 degrees C for various materials used to make wires is shown in Table 102-2. This tells how much resistance is contributed by every meter of wire.

Table 102-2

Material	Resistivity (x10⁻⁸ ohm-cm)
Aluminum	2.8
Copper	1.7
Iron	10
Nichrome	100

Why It Works

For a given wire size, resistance is proportional to the material's resistivity, according to the equation:

$$R = \rho L/A$$

where R is resistance in ohms

ρ is the resistivity in ohm-cm (ρ is the Greek letter "rho")

L is length in centimeters

A is cross-sectional area in centimeters.

Other Things to Try

A less precise, but possibly more fun, approach to this experiment is to use wires cut from food items, such as pickles or fruit, or by forming wires from Play Dough.

The wire can be sliced in sections as it is measured to shorten its length. This approach may require the use of two meters because the ohmmeter may not be stable.

This can be taken a step further by comparing the *resistance* to the *resistivity*. You can get the resistivity by multiplying the resistance by the length of the wire (in cm) and the area of the wire (in cm²). You can get the area of the wire from using a measured or looked-up value for the wire diameter and using the equation:

$$A = \pi r^2$$

where r is the radius of the wire.

The Point

Resistance is a measure of how difficult it is for a given voltage to force electrons through a conductor. It doesn't matter how big or small the piece of conductor. All that matters is the overall effect it has in the electrical circuit.

On the other hand, *resistivity* is a measure of how effective a particular material is in impeding the flow of electrons. Resistivity is the same for any particular material.

Resistance combines the effect of the material's resistivity, as well as its length and cross-section.

Project 103
Storing charge. Capacitors.

The Idea

A *capacitor* is an electronic component that can store an electrical charge. Unlike a battery that stores electrical charge through chemical reactions, the capacitor holds electrons on conductive plates separated by an insulator. Capacitors are present in numerous electronic circuits. They are also gaining attention recently as a possible means of supplementing batteries in electric cars. This experiment explores how capacitors can be charged and discharged.

What You Need

- 1000 μF (micro-Farad) capacitor
- 50 kΩ (kilo-Ohm) resistor (note other capacitor/resistor combinations that can work are listed in Table 103-1)
- DC voltmeter (or multimeter configured as a voltmeter)
- 10-volt DC power supply
- DC ammeter (with 0–1.0 mA range)
- 3 knife switches (SW1, SW2, and SW3)
- jumper wire
- stopwatch
- 2 LEDs

Method

Charging

1. Set up the circuit shown in Figure 103-1. Pay attention to the positive and negative polarity markings, especially if your capacitor has a designated positive side (some do and some don't). Start with all switches open.

2. Close SW2. Leave open SW3.

3. Close SW1 and start the timer.

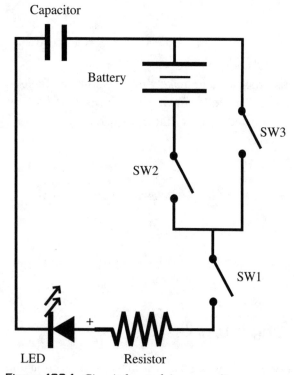

Figure 103-1 *Circuit for studying capacitors.*

4. Record the current in mA every five seconds (this is easier with partners). If you miss a reading, keep going and catch the next five-second interval. Keep going until the current becomes too small to read. If other capacitor/resistor combinations are used, a different time interval than five seconds may be more appropriate.

Discharging

1. When the charging part is complete, open all the switches.

2. Close SW1 and leave SW2 open.

3. Close SW3 and start the timer.

4. As before, record the current in five-second intervals.

Expected Results

With SW2 closed, the capacitor will charge. LED2 will light, but slowly fades as the voltage builds and the current flow decreases. For the 10 kΩ resistor and the 1000 μF capacitor given in the parts list, the charging will be about two-thirds complete in 50 seconds, as shown in Figure 103-2.

With SW3 closed, the capacitor will discharge as indicated in Figure 103-3. After 50 seconds the voltage will have dropped from 10 volts to around 3.7 volts. LED3 will light and will slowly fade as the capacitor discharges.

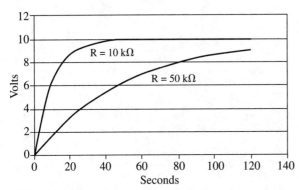

Figure 103-2 *Capacitor voltage versus time for a 1000 μF capacitor charging through 10 kΩ and 50 kΩ resistors.*

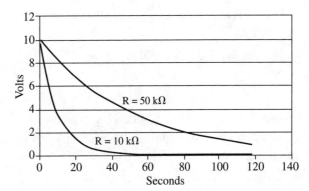

Figure 103-3 *Capacitor voltage versus time for a 1000 μF capacitor discharging through 10 kΩ and 50 kΩ resistors.*

In general, the time to charge or discharge two-thirds of capacity is characterized by the time constant. For a capacitor C (in Farads) and a resistor R (in ohms), the time constant, τ (in seconds), is given by $\tau = RC$. The *time constant* represents the time where the current during charging or the voltage during discharging has decreased by about two-thirds. The following combinations of resistor and capacitor (Table 103-1) give a reasonable time constant of 30 seconds, which gives measurable results in this experiment.

Table 103-1

Resistance	Capacitance
30 ohms	1 F
300 ohms	0.1 F
3000 ohms or 3 kΩ	0.01 F or 10,000 μF
30000 ohms or 30 kΩ	0.001 F or 1000 μF
300000 ohms or 300 kΩ	0.0001 F or 100 μF
3000000 or 3 MΩ	0.00001 F or 100 μF

Why It Works

The current for a charging capacitor is given by $I = I_o e^{-t/RC}$

The voltage for a charging capacitor is given by $V = V_o(1 - e^{-t/RC})$

The current for a discharging capacitor is given by $I = I_0 e^{-t/RC}$

The voltage for a discharging capacitor is given by $V = V_0 e^{-t/RC}$

When t = the time constant, RC, then $e^{-t/RC} = e^{-1} = 0.37$. This mean a discharging capacitor has dropped to about one-third of its original value or has discharged about two-thirds.

Other Things to Try

If you have other resistors and capacitors available, try (small) increases or decreases in values, and then determine how it affects the time to charge and discharge. The previous Table 103-1 gives combinations that result in reasonable time constants and serves as a good starting point. Adjust your measurement interval as needed.

Use a current and voltage sensor that displays these parameters as a function of time on a computer. A combination voltage/current sensor (part number PS-2115) is available from PASCO that displays both parameters simultaneously in DataStudio software.

Make a graph of voltage (or current) versus time for your discharge data with voltage on a linear scale and time on a logarithmic scale. Use an exponential curve fit to an Excel scatter plot to find the argument of the exponent. Compare that with $-1/RC$.

The Point

A capacitor is a device that stores electrical energy. The rate of charging and discharging depends on the size of the capacitor and the resistor it is charging or discharging through. The bigger the capacitor and the resistor, the longer these processes take. The charging and discharging is an exponential function of time that approaches a saturation value.

Project 104
Is the magnetic force more powerful than gravity?

The Idea

In this experiment, you use a magnetic field to defy gravity and hold a magnetic object suspended in the air. You also explore the effectiveness of various materials in shielding the effects of the magnetic field.

What You Need

- powerful permanent magnet
- ring stand with clamp
- paper clip
- 12 inches of (low mass) string
- materials to test as a shield: glass, paper, copper
- piece of tape

Method

1. Secure the magnet to the ring stand, so the most powerful magnetic field is directed downward.

2. Tie the string to the paper clip.

3. Bring the paper clip near the overhead magnet. It should be close enough for the magnetic field to exert a force on the clip, as shown in Figure 104-1.

4. Tape the other end of the string to the table. Using tape enables you to easily make slight adjustments in the string length.

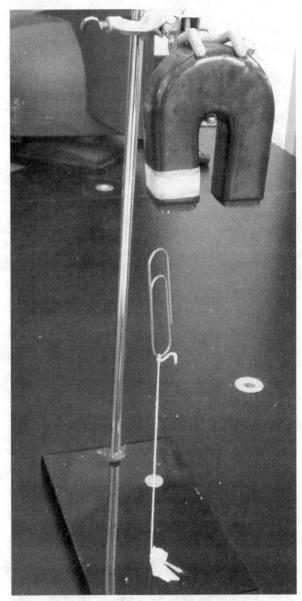

Figure 104-1 *Photo by S. Grabowski.*

5. At this point, if you are going to show this to someone, this would be a good point to have them come into the room.

6. Observe what happens when you bring the paper clip closer, and then further from the magnet.

7. Try blocking the magnetic field by using any of the following potential shielding materials: glass, paper, copper, iron.

Expected Results

The paper clip appears to defy gravity and will be held suspended above the table. Depending on the strength of the magnet, a gap of a few millimeters can be established between the paper clip and the magnet.

The magnetic field can be shown to penetrate through materials, such as glass, paper, wood, or copper. See Figures 104-2 and 104-3.

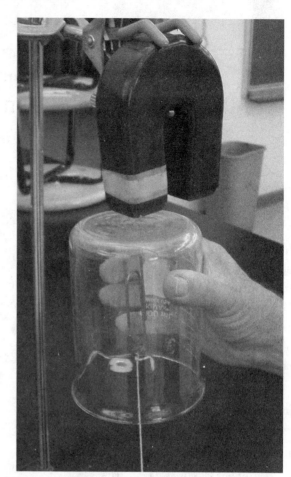

Figure 104-3 *Does a magnetic field pass through an insulator? Photo by S. Grabowski.*

Why It Works

The force exerted by the magnet is greater than the force of gravity.

Other Things to Try

Use a very sensitive spring scale or a force gauge to measure the magnetic force exerted on the paper clip.

The Point

Magnetic fields exert a force on magnetic objects. This force decreases with distance. The magnetic field penetrates both electrical insulators and conductors.

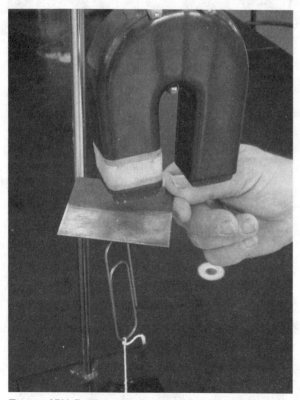

Figure 104-2 *Does a magnetic field pass through a conductor? This should leave little doubt. Photo by S. Grabowski.*

Project 105
Magnetic levitation using induction.
Electromagnetic ring tosser.

The Idea

This is a fun demo with somewhat surprising results. You will use a powerful magnetic field to exert a force on a material that is not normally magnetic. You will generate a circulating current using electromagnetic induction. The result is that the magnetic repulsion causes a metal object to be forcefully thrown into the air.

What You Need

- ring launcher apparatus (Elihu Thomson apparatus). Pictured in Figure 105-1 is PASCO EM-8661.
- copper collar
- aluminum ring
- split aluminum ring
- copper ring
- lead ring
- coil of insulated wire connected in series with a low-wattage light bulb
- AC voltmeter (or multimeter configured as an AC voltmeter)
- tongs
- optional: Pyrex bowl with liquid nitrogen

Method

1. Safety: All wiring to the launcher should be appropriately enclosed and insulated to avoid a potential shock hazard. Without giving away the results of this experiment yet, it should come as no surprise that metal objects will be launched, sometimes at significant velocities. *Make sure no one and nothing of value can be hit by flying rings.* Caution should be exercised when working with low-temperature rings. Avoid constraining any of the collars for any prolonged period, which could result in elevated temperatures and burning hazard. The current should flow through the coil of the launcher for a limited time. Be careful to avoid overheating the coil by flowing current for an excessive time.

2. Slide the aluminum ring over the iron core of the ring launcher apparatus and slide in down over the coil.

3. Activate the launch switch for a few seconds to apply 120V alternating current to the coil.

4. Repeat with the other rings and collars.

5. Connect an AC voltmeter across the two sides of the split ring to measure the current while

Figure 105-1 *Courtesy PASCO.*

the current is flowing in the core of the launcher. Similarly, put a coil of wire around the core. Compare the AC voltage developed in one, two, or multiple coils of wire.

6. Place the coil of wire attached to the light bulb over the launcher coil and apply current to the launcher, as shown in Figure 105-2.

7. Hold the copper ring over the collar for a few seconds with the launcher turned on and feel the collar's temperature. (Be careful not to hold it too long or it may get too hot to handle.)

8. For this next part, make sure the ceiling is high enough. Immerse the rings in liquid nitrogen and activate the launcher.

Expected Results

Closed conductive rings and collars will be throw vertically by the ring tosser, as shown in Figure 105-3. The copper ring is thrown the highest,

followed by aluminum. The copper collar is raised, but it is more sluggish. The split rings are not lifted. An AC voltage of a few millivolts can be measured across the split ring. A ring that is held down while the current is flowing in the coil will heat up significantly. The rings cooled in liquid nitrogen are much more response than their room temperature counterparts. The cooled copper ring will fly the highest and can likely damage a standard 8 to 10 foot (2 meter) ceiling. The light bulb will be illuminated when held over the current-carrying coil.

Why It Works

This is a demonstration of electromagnetic induction based on an apparatus developed by the prolific inventor Elihu Thomson. A constantly changing magnetic field produced by the applied alternating current causes an opposing current,

Figure 105-2 *Induced current causes light to be illuminated. Courtesy PASCO.*

Figure 105-3 *Ring tosser. Courtesy PASCO.*

and voltage in the rings and collars. The generation of an opposing current is an illustration of Lenz's law.

This induced current gives rise to a magnetic field oriented to repel against the field that forms in the ring launcher coils. The repulsion between these magnetic fields causes the ring to be tossed.

Because copper is a better conductor than aluminum or lead, more current flows and the ring is tossed higher. The collars are heavier and are not thrown as far. The split rings do not provide a complete current path, so the induced current does not flow in a complete circuit. The bulb lights because a current is induced in the coil connected to the bulb. The liquid nitrogen reduces the resistance of the rings. With lower resistance, more current can flow. The higher current creates a stronger magnetic field, which launches the ring higher.

Other Things to Try

The current generated in the split ring can be measured by attaching an AC voltmeter or a multimeter configured as an AC voltmeter.

The Point

A current flowing in a conductor produces a magnetic field. A changing magnetic field can induce a current in a conductor. The induced current can then generate a current. These currents according to Lenz's law will always oppose each other.

Project 106
Magnetic levitation using superconductivity.
The Meissner effect.

The Idea

Typically, when the temperature of a conductor is reduced, the resistance is also lowered. We saw in the previous experiment how a magnetic field can cause an object to levitate. For some materials, if we continue to lower their temperature, the resistance continues to drop until it disappears entirely. When this happens, we have what is known as a *superconductor*. Superconductors have amazing properties and are beginning to find their way into practical applications.

Figure 106-1 *Neodymium magnet cube.*

What You Need

- liquid nitrogen
- thin piece of cork (about ¼ inch thick)
- Styrofoam dish (formed by cutting a Styrofoam cup; tThe total height should be about 2 mm)
- cube-shaped neodymium magnet (see Figure 106-1)
- plastic tongs
- superconductor disk consisting of $YBa_2Cu_3O_7$ ceramic (see Figure 106-2)
- optional: video camera or PC cam connected to a TV monitor (to show this to a larger group)
- optional: thermocouple, voltmeter, DC power supply, superconductor coil, superconductor sample with measurement leads attached

Figure 106-2 *$YBa_2Cu_3O_7$ ceramic disk.*

Method

1. Place the $YBa_2Cu_3O_7$ ceramic disk on the table and set the neodymium magnet on top of it to show no repulsive force is occurring at room temperature.

2. Place the cork in the center of the Styrofoam dish.

3. Place the black $YBa_2Cu_3O_7$ ceramic disk on the piece of cork.

4. Carefully pour liquid nitrogen into the Styrofoam dish to partially cover the ceramic disk.

5. The liquid nitrogen will boil for a short while. When the boiling subsides, the disk has sufficiently cooled, as shown in Figure 106-3.

6. Using the plastic tongs, pick up the neodymium magnet. Carefully place the magnet over the ceramic disk.

7. When the magnet is observed to hover over the ceramic disk, use the tongs to give it a spin, as shown in Figure 106-4.

8. Because the parts in this project are small, if the intention is to show this to a larger group, a video camera or PC cam can be used to display this on a monitor.

Expected Results

The magnet is held suspended above the ceramic superconducting material. If the magnet is spun, it continues spinning without noticeable resistance. Eventually, the ceramic will warm up and the superconducting effect will fade.

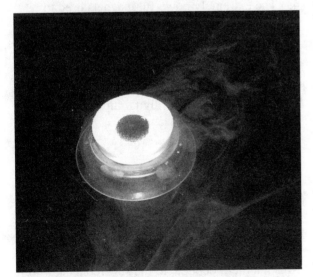

Figure 106-3 *Superconducting disk being brought below Curie temperature.*

Figure 106-4 *Magnetic cube spinning above superconducting ceramic disk.*

Why It Works

Normally, at temperatures above what is known as the critical temperature of a material, the material has some electrical resistance. This means a voltage must be applied across the material to push the electrons through the material. The voltage is needed to drive the electrons through what is like an atomic obstacle course, consisting of other atoms vibrating randomly. As (normal nonsuperconducting) resistors cool down, their resistance gets lower. However, superconductors have zero resistance. Not just lower, but zero! This means the electrons no longer need a voltage to push them. This also means the electrons can move about freely throughout the superconductor without energy losses.

Different materials become superconducting at characteristic temperatures that differ for each material, as shown in the following table:

Material	Type	$T_c(K)$
Zinc	Metal	0.88
Aluminum	Metal	1.19
Tin	Metal	3.72
Mercury	Metal	4.15
$YBa_2Cu_3O_7$	Ceramic	90
$Hg_2Tl_{.8}Ca_2Ba_2Cu_3O$	Ceramic	139

Notice that all the metals listed must be cooled to below 4.2 K. To accomplish this, it is necessary to use liquid helium, which remains liquid up to that temperature, as shown in the following table. In 1987, a breakthrough was achieved by the discovery that $YBa_2Cu_3O_7$ ceramic became superconducting around 90 K (and below), which can be achieved by immersion in liquid nitrogen. This is far less expensive and easier to work with than liquid helium. Scientists are pursuing materials that can be superconducting at temperatures closer to room temperature, which could open the door to its application in many new areas.

Condition	Fahrenheit (°F)	Centigrade (°C)	Kelvin (K)
Absolute zero	−460	−273	0
Boiling point of helium	−452.1	−268.8	4.2
Boiling point of nitrogen	−321	−196	77
Freezing point of water	32	0	273

The actual details of how superconductivity works will take us further into quantum mechanics than I think most readers would care to go. The theory known as BCS theory was named after three American physicists: Bardeen, Cooper, and Schrieffer. (Very) basically, the *BCS theory* describes how electrons are able to more easily navigate through the crystal lattice of matter in a manner that is somewhat analogous to a race car encountering less aerodynamic resistance as it closely follows another car in front of it. As the critical temperature is reached, the electrons are able to go through a material by "tunneling" right through an electrical field in its way. As a result, superconductors have zero resistance. If digital electronics were based on superconductors they would function ten times faster than standard semiconductor electronics.

Magnetic fields can pass through and be present in most materials, including superconductors above their critical temperature. However, as a superconducting material is brought *below* its critical temperature, the magnetic field is forced out in a process known as the Meissner effect, which serves as the basis for the effect we saw here. To enable the magnetic field to be pushed out of the superconductor, it becomes necessary for a counter current to flow in the superconductor. With no resistance, electrical currents are induced in the superconducting ceramic, which, in turn, creates a magnetic field that repels against that of the permanent magnet.

Superconductors are being looked at to address some of the following challenges in technology:

1. Much of the electrical power transmitted throughout the world's electrical power grid is dissipated as resistive heat losses. If superconductors could be used for power transmission and generation, some of the losses could be reduced.

2. Magnetic resonance imaging (MRI) equipment uses extremely powerful magnets to help create detailed images of the body. Superconductors allow stronger magnets to be built.

3. *Maglev* trains use superconductors to help produce powerful magnetic fields that raise trains above the track, enormously eliminating friction.

4. The extremely powerful magnets used to guide beams of subatomic particles in research facilities (such as CERN) use superconductors.

5. Superconductors hold the promise of enabling faster processing of digital information in computers.

Other Things to Try

Some additional experiments include:

Measure the critical temperature

Attach a thermocouple to the superconductor to measure the critical temperature. This is the temperature at which the magnet first begins to levitate about the cooled ceramic disk. The thermocouple should not constrain the magnet from being lifted and should not be immersed in the liquid nitrogen or it will understate the critical temperature.

Resistance versus temperature curve

Basically to do this you will use Ohm's law (resistance = voltage/current) to measure the resistance of the superconductor as its temperature changes. It is easier to measure these changes as the superconductor warms from the initial immersion into liquid nitrogen until it passes through the critical temperature. Two wires for current and two wires for voltage are attached to the superconductor. This is called a *four-point probe*, which eliminates the effect of contact resistance that would be encountered if both voltage and current were measured using a single set of contacts. The contacts can be formed using very high gauge (thin) silver or copper wire and attaching them to the superconductor using silver paint. A sample with contacts attached can be obtained as part of an experiment kit by superconductor experiment suppliers.

The Point

Superconductors are materials that have no electrical resistance when they are brought below their critical temperature. Superconductors can exert of force on a permanent magnet as a result of the Meissner effect, in which a circulating current is established in response to exclusion of the magnetic field. This current generates a magnetic field that repels the permanent magnet. Superconductors have significant technological applications, including MRIs, maglev trains, subatomic particle research, and ultrafast digital electronics.

Project 107
Moving electrons produce a magnetic field. Oersted's experiment. The magnetic field of a current-carrying wire.

The Idea

What causes a magnetic field? In 1820, Hans Oersted discovered that electricity flowing in a wire caused a compass needle to be deflected. This established one of the earliest connections between electricity and magnetism. This investigation re-creates some of the things Oersted did.

What You Need

- DC power supply or battery
- about a meter (a few feet) of insulated wire
- compass, preferably one mounted on a pivot
- ring stand or other support for the wire
- optional: ammeter
- optional: knife switch to complete the circuit. You can activate the circuit simply by completing the final connection to the power supply.

Method

1. Connect one end of the wire to the positive terminal of the power supply (or battery).

2. Route the wire through the support, so a length of about 0.25 m (or about a foot) is running up and down. The current flowing from the positive terminal should flow up through the wire.

3. The other side of the wire should either go to one of the terminals of the switch or be placed ready to attach to the negative terminal. If you are using the knife switch, keep the switch in the open position and attach the other terminal to the negative terminal.

4. Make sure no other magnets are in the immediate vicinity of the apparatus.

5. Place the compass close to the wire. (The compass will point in the direction of the Earth's magnetic field. Imagine a circle formed around the wire when viewed from above. To experience the force generated by the magnetic field of the wire, the compass needle should not be at a tangent to that circle, but it should form some angle to the tangent at that point. This is shown in Figure 107-1.)

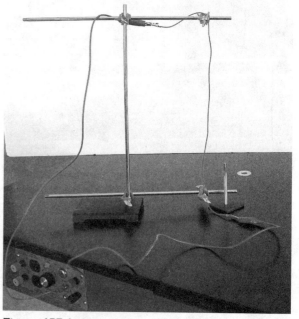

Figure 107-1 *Measuring the effect of current on a magnet.*

6. Complete the circuit (either by closing the switch or by attaching the wire to the negative terminal). If you are using an adjustable power supply, it may be helpful to put an ammeter in series with the power supply to make sure the current flow doesn't go much above 1 amp. This can also let you quantify the effect of increasing current on the strength of the generated magnetic field.

Expected Results

With a current flowing through the wire, the compass is deflected. The magnetic field deflects the compass needle in the direction of the tangent of the circle, as shown in Figure 107-1.

Why It Works

A magnetic field is produced by moving electric charges.

Other Things to Try

Measure the effect of increasing the current and moving the compass further from the wire on the response of the compass needle.

The Point

An electrical current flowing in a conductor produces a magnetic field.

Project 108
Faraday's experiment. Current generated by a magnet.

The Idea

Michael Faraday discovered that a moving magnetic field causes an electrical current to flow in a wire. Most of the electricity generated throughout the world today is based upon this historic discovery. Power plants routinely convert mechanical energy into electrical energy. This experiment explores the physical principle that makes this possible.

What You Need

- bar magnet
- coil of insulated wire—the more coils, the more pronounced the effect. Thin "magnet" wire insulated with clear or colored enamel can work fine.
- galvanometer (a very sensitive ammeter)

Method

1. Wrap the insulated wire in a coil around a cylindrical cardboard form. Use the smallest diameter that will enable the bar magnet to pass through. Prewound coils are available.

2. Connect the two ends of the wire to the positive and negative terminals of the galvanometer.

3. Predict what you think will happen if the magnetic is placed inside the coil. Try it and observe the response on the galvanometer.

4. Move the magnet back and forth in the coil. Observe the deflection on the galvanometer.

5. Move the magnet back and forth outside the coil and observe the effect. What happens if the coil moves while the magnet is stationary?

6. If it is possible to increase or decrease the number of coils, you can evaluate its effect on the amount of current that can be generated.

7. Based on your observations, describe how a magnet can produce an electric current.

The apparatus is shown in Figure 108-1.

Expected Results

A magnet produces a current in a wire only when the magnet is moving. A stationary magnet will not generate a current.

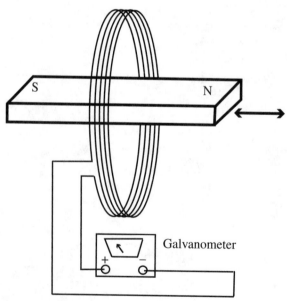

Figure 108-1 *A magnet moving through a coil of wire.*

The faster the relative motion between the magnet and the coil, the greater the current. The larger the number of coils—with all else equal—the greater the current. A more powerful magnet produces a greater current.

Why It Works

A magnetic field itself does not produce an electric current. A changing magnetic field is required to produce an electrical current. This is addressed in mathematical detail by Maxwell's laws for those who want to pursue it further.

Other Things to Try

A nice way to display these results is to attach the coil to a voltage sensor and use this to generate a graph of voltage versus time. Moving the magnet back and forth in the coil results in an alternating current (AC). If you hang the magnet on a spring and have it oscillate up and down in the coil, you will have a simple model of an AC generator.

A good follow-up is to investigate a model electrical generator, which can also generate a similar AC current.

The Point

The significance of this project is to show how mechanical energy is converted into electrical energy. The key parts of an electrical generator are a magnet and a coil of wire. Electricity flows when the coil and magnet move relative to each other.

Project 109
If copper is not magnetic, how can it affect a falling magnet? Lenz's law.

The Idea

When a magnet moves near a conductor, electrical currents can be produced in the conductor. When currents circulate in a piece of bulk material, rather than a wire that forms a complete circuit, the currents are called *eddy currents*. These eddy currents are more like water swirling in a tide pool rather than flowing in a stream. Eddy currents can reduce the efficiency of electrical devices, such as transformers, because the circulating current results in a loss of power. Eddy currents have several uses, which include magnetic damping of sensitive meters, and they are used in magnetic braking for rapid transit trains. This project explores an aspect of eddy currents called Lenz's law.

What You Need

- 3 neodymium disc magnets
- copper pipe whose diameter is slightly larger than the magnet
- plastic pipe of similar dimensions

Method

1. Hold both pipes vertically.
2. Check the magnetic attraction between the magnets and each of the pipes.
3. Position the magnet over the open top of each of the pipes.
4. Drop the magnet through the pipes at the same time.
5. Compare how fast the magnets fall through each of the pipes, as shown in Figures 109-1 and 109-2.

Expected Results

The magnet falls through the plastic pipe faster than through the copper pipe.

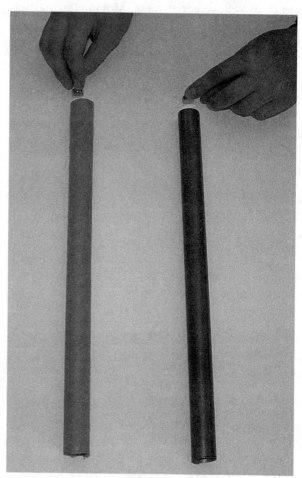

Figure 109-1 *Both magnets are dropped at the same time. The non-conductive tube is on the left.*

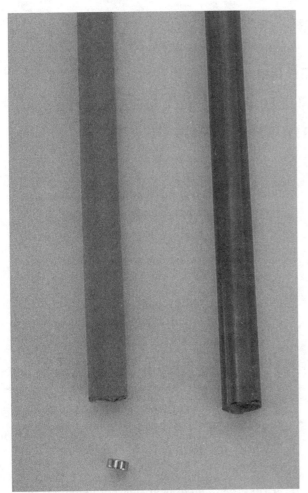

Figure 109-2 *The magnet going through the non-conductive tube (on the left) emerges first.*

Why It Works

When a magnet moves with respect to a conductor (such as copper), it creates (induces) an electric current in the conductor. This current, in turn, produces a magnetic field. According to Lenz's law, this magnetic field will be aligned in such a way that it is pointed in the opposite direction as the magnetic field that originally produced it. This results in an attractive force between the magnet and the copper pipe in which a current is induced by the falling magnet. Because the plastic is not conductive, no magnetic field is produced to create the Lenz effect.

Other Things to Try

One other way to do this is to use disc magnets with holes in their centers. Then, the magnets are placed over copper, plastic, and iron rods whose diameter is slightly smaller than the inner diameter of the hole in the magnet.

An aluminum disc is not magnetic. If the disc is spun and a strong magnet is brought near, eddy currents are produced whose effect is to slow the spinning disc. This is known as *magnetic braking*, and it is shown in Figures 109-3 and Figure 109-4.

The Point

Lenz's law describes how eddy currents are formed in a nonmagnetic material. These eddy currents interact with a magnet in a way that opposes the effect of that magnetic field.

Figure 109-3 *A neodymium magnet is brought near a spinning non-conducting plate.*

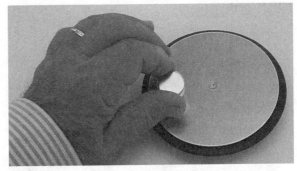

Figure 109-4 *Magnetic braking (and not friction) stops the plate.*

Project 110
Effect of a magnet on an electron beam.
The right-hand rule for magnetic force.

The Idea

One of the great accomplishments of physics in the twentieth century was the discovery of the electron by J. J. Thomson. In this project, you revisit some of the steps that led Thomson to his discovery.

What You Need

- cathode ray tube (CRT)—this can be any tube-type video monitor (TV or computer). If you have access to an oscilloscope, you can use that. Liquid crystal display (LCD) or plasma screen displays do not have electron beams and will not work here. A stand-alone CRT (such as a Crooke's tube available from scientific supply companies excited by an induction coil) can be used.
- strong magnet
- your right hand
- mini Post-it notes

Method

1. CAUTION: If you are using a high-voltage induction coil to produce the electron beam in the CRT, be very careful not to touch the exposed terminals while it is operating. Be certain you *exactly* follow the manufacturer's instructions for using this equipment.

2. Careful: It is typically not recommended to expose TV or computer screens to magnetic fields. Excessive or prolonged exposure can cause unintended damage.

3. Turn on the CRT.

4 Make any adjustments needed to focus the electron beam.

   - For an oscilloscope, disable the vertical and horizontal sweeps, leaving a single dot at the focus

   - With the computer, find a point of focus, including text or a period on the screen

   - With the TV, find a stationary image to focus on

5. Identify the north pole of your electron magnet.

6. Bring the magnet near (but not touching) the CRT screen.

7. Observe the effect of the magnet on the electron beam.

Expected Results

The magnetic field causes the electron beam to bend.

If the beam is moving *to your right* and the magnetic field, north to south, is pointed *inward* across the tube, the electron beam will be bent down. See Figure 110-1. (Note that since electrons have a negative charge, this is the *opposite* direction specified by *the right-hand rule* for positive charges.)

Figure 110-2 shows an electron beam moving from left to right in a cathode ray tube. In the

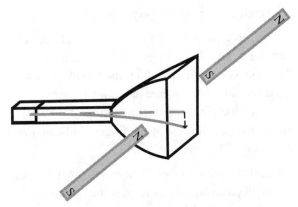

Figure 110-1 *Effect of a magnetic field on an electron beam.*

Figure 110-3 *Cathode ray tube with magnetic field front to back.*

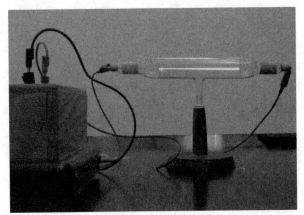

Figure 110-2 *Cathode ray tube. Photo by S. Grabowski.*

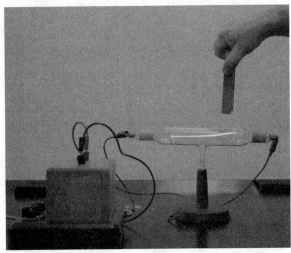

Figure 110-4 *Cathode ray tube with magnetic field back to front. Photo by S. Grabowski.*

absence of a magnetic field, the beam is horizontal.

If a magnetic field is placed across the beam (with the north pole indicated by tape in front), the beam is deflected downward as shown (by Steve Grabowski) in Figure 110-3.

If the magnetic field is reversed (so that the north pole is in the back) the beam is deflected upward as shown in Figure 110-4.

Why It Works

A magnetic field does not exert a force on a stationary electron. However, a magnetic field does produce a force on a *moving* electron. The electron's motion, the magnetic field, and the force are *all* at right angles to each other.

Other Things to Try

The direction of the force on the electron follows the right-hand rule. If your index finger points in the direction of the beam, and your other three fingers point in the direction of the magnetic field (north to south), then your thumb shows the direction of a force on a positive particle (or the opposite direction for a negative particle, such as an electron). You can label the fingers of your right hand with Post-it notes to help keep track of the electron motion, the field, and the resulting force. (This is why Post-it notes are on the What You Need list.)

You can take this a step further. Thomson applied an electric field to deflect the electron beam by a measurable amount. Then, by applying a magnetic field at a right angle to the electric field, Thomson was able to determine the charge to the mass ratio of the electron. This can be performed using commercially available equipment.

The Point

Properties of the electron can be determined by observing its behavior in electric and magnetic fields.

The electron is negatively charged.

The force produced by a magnetic field on a moving beam of electrons can be described by *the right-hand rule*, in which the thumb indicates the direction of the force, the index finger indicates the direction of the motion of the electrons, and the rest of the fingers indicate the direction of the magnetic field.

Careful analysis of electric and magnetic fields on an electron beam determines the charge to mass ratio of the electron.

Project 111
What is the shape of a magnetic field?

The Idea

You cannot see a magnetic field. But you can define the shape of the field by measuring its effects. In this project, you trace the shape of the magnetic field created by various arrangements of permanent magnets.

What You Need

- 2 bar magnets
- U-magnet
- several sheets of paper
- iron filings

Method

1. Lay the bar magnet on the table.
2. Place the sheet of paper over the magnet.
3. Trace the outline of the magnet, showing the north and south poles.
4. Evenly sprinkle iron filings over the paper. Distribute the filings so the shape of the pattern on all sides of the magnet is delineated by the iron filings.
5. Repeat with the following cases.
6. The iron filings can be easily poured back into the container. If they come into direct contact with the magnet, it is much harder to clean up.
   - Two north poles facing each other
   - A north and a south pole facing each other
   - A horseshoe magnet
   - Any other shape—your choice

Expected Results

The electric field surrounding a bar magnet follows lines that go from the north pole to the south pole, as shown in Figure 111-1.

With a north pole directly opposite a south pole, the lines of force are directed from the north pole to the south pole, as shown in Figure 111-2.

With two north poles facing each other, the electric field is directed away from each of the poles. Lines of force can be seen directed perpendicular to each of the two magnets, as shown in Figure 111-3.

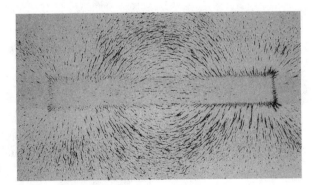

Figure 111-1 *Bar magnet.*

Figure 111-2 *North pole opposite south pole.*

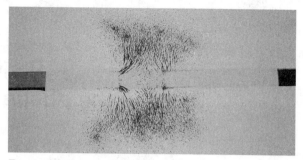

Figure 111-3 *North pole opposite another north pole.*

Why It Works

The magnetic field causes ferromagnetic materials, such as iron filings, to align with the field lines.

Other Things to Try

A higher tech approach would be to use a magnetic field sensor to map out the shape of these magnetic fields.

The Point

Magnetic fields show the force a magnet would exert on the north pole of another magnet. Magnetic fields point from north to south. Magnetic fields point away from north poles (opposites repel) and toward south poles (likes attract).

Project 112
What happens to a current-carrying wire in a magnetic field?

The Idea

The heart and soul of an electrical motor is movement created when magnets repel each other. The discovery by Michael Faraday that a force is produced when current flows through a magnetic field was a groundbreaking discovery, which paved the way for the eventual development of the electric motor.

What You Need

- powerful horseshoe magnet
- DC voltage source, such as an adjustable power supply, a motorcycle battery, or a car battery
- about 1 meter (a few feet) of insulated wire
- ring stand with clamps to position the wire

Method

1. Set the horseshoe magnet on the table.
2. Position the wire midway between the north and south poles of the magnet. The wire should run perpendicular to the two ends of the magnet and it should be able to move.
3. Attach one end of the wire to the negative terminal of the power supply, as shown in Figure 112-1.
4. Briefly touch the other end of the wire to the positive terminal and observe the wire passing between the poles of the magnet. If you are using an adjustable power supply,

start with a lower-current output setting and slowly increase it until the wire responds. Because there is no resistance besides the little resistance the wire offers, a high current may flow that may cause a fuse or circuit breaker to blow.

Expected Results

With current flowing in the wire, the magnet pushes the wire away from the magnet.

Why It Works

A force is exerted on a moving charge (or current) in a wire moving perpendicular to a magnetic field.

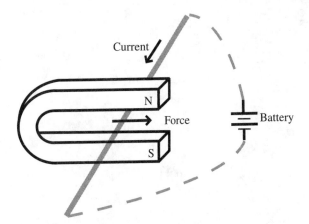

Figure 112-1 *Force between two current-carrying wires.*

Other Things to Try

Repeat this with a tight coil of the wire in the magnetic field. Compare with the response of an uncoiled wire.

The Point

A magnetic field exerts a force on a current-carrying wire. The direction of that force depends on the direction of current flow and the orientation of the magnetic field according to the right-hand rule.

Project 113
A no-frills motor.

The Idea

In this project you will build a very basic electric motor. This project has been broken down into a number of steps for clarity, but the overall device is simple. It's like assembling a barbeque grill—the first time you do it may take a little longer, but once you get the overall idea, it gets easier each time. It takes just a few minutes to build, but it can keep running until the battery runs out.

What You Need

- C or D cell battery
- ceramic disc magnet
- 1 meter of enamel-coated (thin) 22 American Wire Gauge (AWG) wire. It is much easier to work with red or green coated enamel, rather than clear-coated
- 2 paper clips or a few inches of at least 20 AWG wire

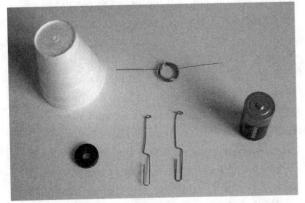

Figure 113-1 *Parts to assemble a "no-frills" motor.*

- electrical tape
- optional: battery holder, Styrofoam cup

Method

1. Wind about 25–30 turns of the 22 gauge wire around a cylindrical coil form, such as a ballpoint pen or a small AAA battery.

2. Leave a few of inches of wire free at each end.

3. Pull the coil off the form you wound it around. Be careful and hold the wire, so it doesn't spring out of shape (it doesn't have to be perfect to work).

4. Weave each of the ends of the wire around the coil a few times to hold the coil together. This becomes the armature of the motor. If you prefer, you can also use tape to help keep the coil together.

5. The ends of the wire should be placed in a straight line to make a good axle. It can help if you double back, so the end sections consist of more than one thickness of wire. It also helps to overwrap the ends with the last segment of wire or tape.

6. Using a utility knife, remove the insulation from the top half of the 22 gauge wire at both ends. You can also use sandpaper to do this. If the wire is coated with a colored layer, you can see when the insulation has been removed. If the wire is clear-coated, you must be more careful and keep track of where you are removing the insulation. Do not remove the insulation from the bottom half of the wire.

7. When you finish, the side with the insulation removed must remain facing up on both ends.

8. Make an armature support by first forming the paper clips into a loop. If you are using wire, remove insulation from each of two 1½-inch sections of wire and form a small circular loop about 1 millimeter in diameter in the center of each wire. A nail can serve as a good form to wrap the loop.

9. Bend the wires to form a shape like a wishbone with the wire ends separated by a few millimeters.

10. The two ends of the coils should easily fit into the armature supports and should be able to turn freely.

11. Secure the armature support wires to the battery holder with tape.

12. Establish electrical contact between the armature supports and the positive and negative terminals of the battery holder. You may need to use jumper wires to do this.

13. Insert the ends of the armature (coil) into the holes of the armature supports. The armature supports should be spaced far enough apart so the coil is supported at both ends.

14. Tape the battery to the top of the cup or insert the battery into the holder.

15. Attach the magnet to the top of the battery holder just underneath the coil. Use tape or Velcro to do this. Make sure the coil can still spin easily and that it is just above the magnet. It may be necessary to raise or lower the armature supports to attain the correct height above the magnet.

16. Spin the armature gently to get the motor started. If it doesn't start spinning, try spinning it in the other direction. It will only spin in one direction.

This may sound like a lot of steps, but it is very simple, as shown by Figure 113-2, which shows what this motor looks like when it is all assembled.

Figure 113-2 *Basic DC motor.*

Expected Results

The motor should keep turning in one direction.

If it does not run, check all electrical connections. Be sure one support touches the negative end of the battery and the other support touches the positive end. Be sure the armature can spin freely. It is essential that the insulation be removed from only one-half of the turns and the uninsulated side of the wire is facing the same direction.

Why It Works

The basic concept of a motor is the repulsion of two magnetic fields, resulting in a repetitive turning motion. One magnet is a permanent magnet. The other is an electromagnet formed by a coil of wire through which an electrical current is flowing. The trick is only to have the magnetic fields repel, but not attract. If we had taken the insulation off the top and bottom sides of the enamel-coated wire used for the coil, the motor would go no more than one-half turn, and then stop as the coil and permanent magnet attracted each other. By leaving the insulation on the bottom halves of the coil wires, no current flows through the circuit at a time when the magnets would attract. In our case, the momentum of the coil keeps it rotating until the uninsulated wires emerge just in time for the permanent magnet to repel the coil and rotate through another cycle. Other motor designs have what is called a *split commutator*, which goes one step better by changing the direction of the current flowing through the wire, so the magnets are always repulsive.

Other Things to Try

Double the spinning power by constructing a split-ring commutator. Try this by making the following modifications to the simple motor concept previously described:

1. Insulate the *top* half of the loop of one armature support and insulate the *bottom* half of the other armature loop.

2. Starting with insulated enamel-coated ends of the 22 gauge wire, remove the insulation from the *top* on one side and the *bottom* on the other side.

3. Once you start the motor, the contacts have been set up to make sure current flows through the coil at a time and in a direction that results in a continuous repulsive force.

The Point

A motor consists of the following fundamental components illustrated in this project: These include a permanent magnet and an electromagnet that receives DC current only during those portions of its cycle when it will be repelled by the permanent magnet.

Project 114
Magnetic accelerator.

The Idea

This is a simple experiment with a very unexpected outcome. A steel ball is rolling in a track drawn by a magnet. The seemingly gentle force produces a powerful acceleration that propels the ball at high velocity. The results are quite amazing and provide an interesting insight into the nature of linear momentum, as well as magnetic fields.

What You Need

- 4 stainless steel balls
- 1 neodeum cylindrical magnet
- track to guide the steel balls (a grooved mounting bracket for curtains works well, and it has the added advantage of providing a handy end "bumper")

Method

1. Place the steel balls in the track.
2. Group three balls together.
3. Roll the fourth ball toward the other three.
4. Notice what happens. (This is not the surprising part, but it establishes a baseline of expectation.)
5. Place three balls on the track. Then, place the neodymium magnet to the right of the three balls.
6. Roll the fourth ball from the right side of the magnet with about the same speed as the ball in Number 3.

Expected Results

Without the magnet, the incoming steel ball stops and knocks out another ball. The dislodged ball continues with the same velocity of the incoming ball. This is the familiar case of conservation of momentum during an elastic collision, as shown in Figure 114-1.

With the magnet in place, a single ball is also knocked out, as shown in Figure 114-2. However, the ball that is knocked out surprisingly moves at turbo speed—much faster than the velocity of the incoming ball. The magnet increases the velocity of the incoming ball. This much higher momentum at the last instant is imparted to the outgoing ball, which shoots off at a surprisingly higher speed.

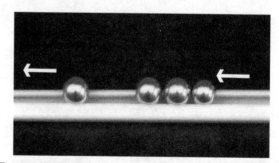

Figure 114-1 *The incoming ball dislodges one ball that exits with the same velocity as the incoming ball.*

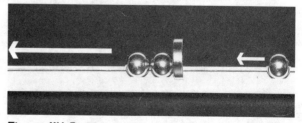

Figure 114-2 *The magnet dramatically increases the momentum of the ball at the last minute.*

Why It Works

In both cases, linear momentum is conserved. With the magnet, the incoming ball is accelerated and achieves a very high instantaneous velocity just before it hits the magnet. Conservation of momentum requires that the outgoing ball moves at that same high velocity.

Linear momentum is always conserved if no force is doing work. In physics, work is force applied over a distance. A principle of physics called the *work-energy theorem* states that if a force is exerted over a distance, the kinetic energy of an object (and, as a result, its velocity) changes. In this case, a magnetic force is doing work, which accelerates the steel ball. Because the magnetic force increases as the ball approaches the magnet, the speed picks up at an even greater rate than a constant force.

Other Things to Try

Repeat with other combinations of balls on either side of the magnet.

The Point

Linear momentum is the same before and after a collision. Because the steel ball is accelerated rapidly by the magnet, the velocity of the ball (and its momentum) is very high just before the collision. Conservation of linear momentum requires the velocity of the ball after the collision also be very high.

Project 115
Alternating current.

The Idea

In this section, you explore some of the basic aspects of AC current.

The electrical power we get from a battery is called direct current (DC). A 9-volt battery produces a voltage of 9 volts, which doesn't change until the battery is used up. The electricity we get delivered from the electrical power company from the wall socket is AC (alternating current). This is different than a battery because the voltage and current coming from our wall sockets is *continuously* changing. The voltage reverses direction 60 times every second in North America (and 50 times each second in most of Europe and much of Asia).

AC is how electricity is distributed throughout the world's power grid. Sometimes DC needs to be converted to AC, such as solar electric panels used to provide power for an electrical utility. Sometime AC needs to be changed to DC at a different voltage, such as is done in cell phone battery chargers.

What You Need

Displaying an alternating current

- waveform generator and an oscilloscope
- connector for the oscilloscope (consisting of a BNC connector with two wire leads attached)
- diode
- alternative: a source of sound, a microphone, and a computer-based, sound-card oscilloscope. CAUTION: Sound card

oscilloscopes can handle *only low-voltage inputs*, such as from microphones. Attempting to use a sound-card oscilloscope for larger electrical signal may damage your sound card. A high-impedance circuit that will enable using a sound-card oscilloscope for higher voltages can be found at www.geocities.com/~uWezi/electronics/projects/soundcard_osci.html.

Building a transformer

- 2-foot length of insulated wire
- 4-foot length of insulated wire
- large iron nail
- AC power supply, waveform generator, or keyboard output
- 2 AC voltmeters (or multimeters configured as an AC voltmeter)

Method

What a 60-cycle AC signal sounds like

1. If you have an adjustable AC power supply, attach one of the terminals of the speaker to the positive terminal of the AC power supply and the other speaker terminal to the negative terminal of the power supply.

2. Slowly turn up the voltage and you will start to hear the characteristic 60-cycle hum coming from the speaker. This may be a familiar sound to rock musicians working with preowned PA systems, which often leaks into audio systems.

What an 60-cycle AC signal looks like

1. Connect the positive and negative terminals of the AC power supply to a 1000 ohm resistor.

2. Attach the two wire leads of the oscilloscope input to the two ends of the resistor. (Do *not* use a PC-based oscilloscope, which we used in other experiments, unless you have a special circuit to adapt the AC signal for this purpose.)

3. Turn on the AC power supply with just enough voltage to produce a display on the oscilloscope.

4. Adjust the amplitude, time sweep, and, if necessary, trigger setting to display the AC signal on the oscilloscope screen. Figure 115-1 shows how the electrical components are connected to make this measurement. (The diode used in the next set of steps is shown connected.)

What a diode does to an alternating current

1. Remove one of the connections to the power supply.

2. Attach a diode in the circuit going from the power supply through the resistor.

3. Turn on the AC power supply.

4. Reattach the leads from the oscilloscope to the ends of the resistor.

5. Display the AC signal on the oscilloscope screen.

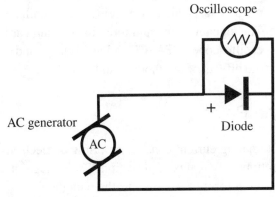

Figure 115-1

6. Turn down the AC power supply.

7. Remove the diode. Reverse the direction of the lead and reattach the diode in the circuit.

8. With the AC power supply turned on, observe how the signal changes.

Building a transformer

1. Wind the 2-foot section of wire around the nail. Leave approximately 6-inch lengths of wire at each end, with about ¾ of the insulation removed from the ends of the wire. Keep track of how many turns you apply.

2. Do the same with the 4-foot section of wire. There should be twice as many turns on this section.

3. Attach the positive and negative of an AC power supply to the two leads of the 2-foot section of wire. (We can call this the primary coil.)

4. Attach the two ends of an AC voltmeter to the points of contact between the power supply and the 2-foot section of the transformer wire.

5. Attach the other AC voltmeter to the two leads of the 4-foot section of wire. What do you read?

6. If you have a DC power supply available, apply a similar voltage to the primary windings. How is the voltage of the secondary affected?

Expected Results

A 60-cycle AC signal is displayed on an oscilloscope with a full wavelength repeating every 0.017 seconds. An AC signal has the form shown in Figure 115-2.

Inserting the diode in the circuit results in only one-half of the waveform flowing in the circuit. This means only the positive (or negative) half of the cycle is displayed, as shown in Figure 115-3 for a diode placed in one direction, or as in Figure 115-4 for a diode placed in the other direction.

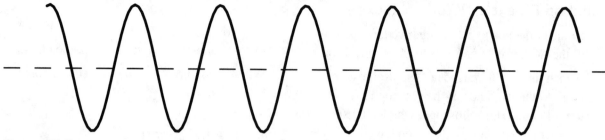

Figure 115-2 *Alternating current waveform.*

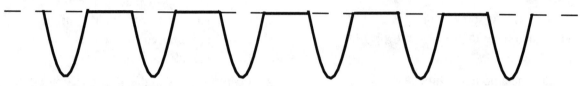

Figure 115-3 *Alternating current with a diode.*

Figure 115-4 *Alternating current with a diode facing the other way.*

Why It Works

Alternating current is constantly changing direction.

A diode is a device that passes the current in only one direction.

A transformer changes the AC voltage of an incoming signal based on the ratio of turns between the input and output sides of a transformer. A transformer only lets AC current through, but it will not pass DC current.

The ratio of the primary (in) to the secondary (out) voltage of a transformer is the ratio of the turns of the secondary to the primary. This is given by the equation $V_p/V_s = N_p/N_s$ where V represents the voltage, N the number of windings, p the primary, and s the secondary windings.

Other Things to Try

If you don't have a stand-alone oscilloscope, here are some other options:

1. Build an adapter for the sound card oscilloscope.

2. Use an audible tone, such as from an electronic synthesizer keyboard, to produce a signal that is compatible with a sound card oscilloscope.

3. You can also generate an AC signal using a magnet suspended by a spring over a coil. The signal can be monitored by a sound card oscilloscope or PASCO voltage sensor, and the effects of the diodes can be studied.

The Point

Alternating current consists of a flow of electrons continuously reversing direction. The voltage of a common form of AC follows the rising and falling pattern of a sine wave.

Project 116
The diode. An electronic one-way valve.

The Idea

A *diode* is an electronic device that lets current flow in only one direction. Diodes are found in electronic circuits and form the basis for more complicated devices, such as transistors and integrated circuits. LEDs (light-emitting diodes) and solar cells are diodes.

Unlike the resistors we studied in previous experiments, diodes do not follow Ohm's law. They are called nonlinear devices, which gives them properties that are useful in a wide variety of electronic applications.

What You Need

- diode
- DC power supply
- voltmeter
- ammeter
- jumper wires

Method

1. Set up the circuit, as shown in Figure 116-1. This consists of the positive terminal of the power supply connected to the positive end of the diode (identified by the longer of the two leads). The ammeter is connected in series with the diode and, together, they are attached to the power supply. The voltmeter is connected to the two terminals of the diode.

2. Start with the power supply at the lowest level and make sure the voltage and current meters read zero.

3. Very slowly, walk the voltage up, taking current and voltage readings at each step. Continue until the current suddenly goes up significantly higher than previous levels. Do not allow too much current to flow or the diode can be damaged.

Expected Results

The relationship between voltage and current is not linear.

As the voltage increases, a threshold is reached where a small increase in voltage results in a huge increase in current.

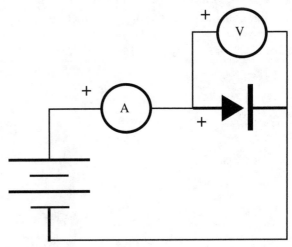

Figure 116-1 *Diode test circuit.*

This relationship between current and voltage is exponential.

Why It Works

The current that passes through a diode is related to the voltage applied across it by the diode equation:

$$I = I_o e^{qV/kT}$$

where I is the current and V is the voltage (q and k are constants, T is the diode temperature, and I_o is a property of the diode).

This equation shows that the current increases exponentially as the voltage is increased. At first the change is slow. But after about 0.7 V, the diode offers little resistance to the flow of current.

Other Things to Try

Current-voltage characteristics

Plot the current versus voltage on a linear plot. Its shape is exponential with a "knee" around 0.7 V defining a region where a small increase in voltage causes a very large increase in current. If you plot the log of the current versus the voltage, you should get a straight line at least over a major part of the data range. If voltage is a logarithmic function of current, current is an exponential function of voltage. Plotting this confirms the nonlinear behavior of the diode characteristic.

Voltage in the reverse direction

What happens if you swap the two leads of the diode? This applies a voltage in the opposite direction as in the previous case. As a one-way valve, the diode does not allow any measurable current to flow in the reverse direction. If you try to force the issue and continue to increase the voltage (going the wrong way), you may (depending on the diode) reach a condition called the *breakdown voltage*. When this happens, the opposition to the current flow breaks down and the diode allows the current to flow. In many diodes, this is a reversible condition, which can be used to establish a set voltage level in a circuit. Going into breakdown may damage some diodes, so be careful if you try to measure this in your circuit.

The Point

A diode is a nonlinear device. A small increase in voltage produces a large increase in current, which grows exponentially with voltage.

The Earth

Project 117
Measuring the Earth's magnetic field.

The Idea

The Earth has a magnetic field that goes from the South Pole to the North Pole. The *magnetic* South Pole is actually close to the *geographic* North Pole. We can measure how strong the horizontal component of the Earth's magnetic field is by comparing its effect to that of a magnetic field produced by the current flowing in a coil of wire.

What You Need

- insulated wire several meters in length
- compass, preferably one mounted on a low-friction pivot
- ruler
- protractor
- cylindrical shape to wrap the coil (The diameter of the shape depends on the length of the compass needle. The diameter of the coil needs to be larger than the length of the compass needle.)
- ring stand or other support to hold the coil of wire

- DC power supply capable of current in the range of 1.0 amp or higher
- DC-ammeter or a multimeter configured as an ammeter in the 0–10 A range
- room with nonferrous tables and free of stray magnetic fields

Method

1. Set up the compass. Make sure it is free-spinning and *pointing to the north*. Metal desks containing iron or steel may interfere with this. Also, motors or loudspeakers may have significant magnetic fields that could affect the outcome of this measurement.

2. Form a coil of 15 turns using the cylindrical shape to form the coil. (For a small hand-held compass, a 1½ inch diameter pipe is a good form. For the pivot type compass, a soup can or coffee can is more appropriate.) After the coil is formed, withdraw the object used to wind the coil. Leave some wire at the start and end of the coil to allow it to be connected into a circuit.

3. Support the coil using a ring stand or other support. The coil is oriented vertically with

the plane of the coil facing east and west. The compass should be contained inside the plane of the coil, as shown in Figure 117-1. A top view of this is shown in Figure 117-2 for clarity. Notice the ends of the compass points to the turns of the coil.

4. Make sure the DC power supply is turned off and the ammeter is set to read currents in the range of 1–10 amps.

5. After stripping the insulation from the ends of the coil, attach one end to the positive terminal of the ammeter and the other end to the negative terminal of the DC power supply. You can use jumper wires or attach the coil directly. Refer to Figures 117-1 and 117-2 for the appropriate connections.

6. Complete the electrical circuit by connecting the negative terminal of the ammeter to the positive terminal of the DC power supply.

7. Place the protractor so the zero degree line ins aligned with the direction the compass exposed only to the Earth's magnetic field.

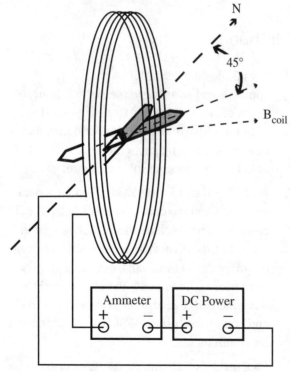

Figure 117-1 *Set up for measuring the Earth's magnetic field (side view).*

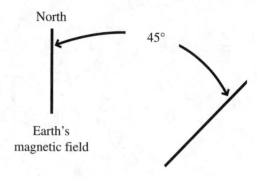

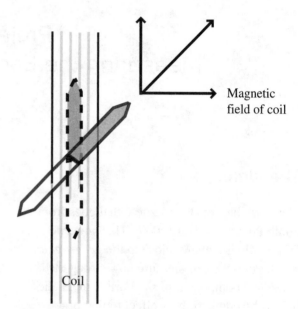

Figure 117-2 *Set up for measuring the Earth's magnetic field (top view).*

8. Slowly and carefully turn on the DC power supply. Increase the current reading on the ammeter until the compass needle deflects 45 degrees from its starting position.

9. At this point, the horizontal component of the Earth's magnetic field is balanced by and equal to the magnetic field of the coil.

The Earth's magnetic field is:

$$\frac{15 \text{ coils} \times (1.26 \times 10^{-6}) \times (\text{the current in amps})}{2 \times \text{the radius in meters}}$$

(1.26×10^{-6} is the same as 0.0000126 and you can multiply inches by 0.00254 to get meters.)

Expected Results

The Earth's magnetic field varies with location, but it is in the ballpark of about 5 micro Teslas or 5 µT or 5×10^{-7} T. The following table summarizes the results for different latitudes, and it includes the scientific notation and decimal forms that are equivalent.

	Scientific notation	Decimal
Near the poles	6.0×10^{-5} Teslas	0.000060 Teslas
Northern US (Michigan)	5.8×10^{-5} Teslas	0.000058 Teslas
Southern US (Florida)	4.8×10^{-5} Teslas	0.000048 Teslas
Near the equator	3.0×10^{-5} Teslas	0.000030 Teslas

Why It Works

The magnetic field of the coil is perpendicular to the plane of the coil. In this experiment, the magnetic field of the coil is perpendicular to the horizontal component of the magnetic field of the Earth. When the coil's magnetic field just equals the horizontal component of the Earth's magnetic field, the resultant points at a 45-degree angle between the two. When this occurs, the magnetic field of the Earth is given by that of the coil according to the equation:

$$B = \frac{N\mu_o I}{2R}$$

where B is the coil's magnetic field in Teslas

N is the number of turns in the coil

μ_o is a measure of how strong a magnetic field is produced by a given current, called the *permeability of free space*, and equals 1.26×10^{-6} Teslas

I is the current in amps

R is the radius of the coil in meters.

As an example: A 15-turn coil that is 2 inches in diameter (or 0.051 meters) requires a current of 0.28 amps to turn the compass 45 degrees.

The magnetic field is:

B = (15 turns $\times$ 1.26 $\times$ 10^{-6} T-m/A $\times$ 0.28A) / (2 $\times$ 0.051m)

= 0.00000052 T or 5.2×10^{-5} T

This is in the ballpark of the expected range for the Earth's magnetic field for middle latitudes.

Other Things to Try

The previously measured value is the horizontal component of the Earth's magnetic field. Near the equator, the Earth's magnetic field is all horizontal. As you approach the poles, the direction of the magnetic field with respect to the Earth's surface increases. The angle the field makes with the Earth's surface can be measured using a compass that is free to rotate in the vertical plane. The total field (or the overall field strength vector) at that location can be determined from:

total field (Teslas) = horizontal component (Teslas) / cosine (angle to horizontal)

As with many experiments, it is comforting to know that the effect we intend to measure is, in fact, what our experimental results are giving us. One way to increase confidence in our results is to repeat it under different condition and verify we have the same outcome. According to our model, it should not matter how many coils we have. Repeating the measurement to see how much current is needed to turn the compass 45 degrees using 5, 10, or 20 coils should give a consistent result as the measurement described about using 15 coils.

The Point

The magnetic field of the Earth can be measured by balancing it with a known magnetic field. If that magnetic field is at right angles to the horizontal component of the Earth's magnetic field, the compass will point in a new direction that is 45 degrees from the original position. Away from the equator, the Earth's magnetic field is at an angle to the horizontal, which can be measured. The overall magnetic field will be slightly higher than the horizontal component.

Project 118
Weighing the Earth.

The Idea

Shortly after Sputnik was launched by the former USSR, President Dwight David Eisenhower, asked his generals to tell him, *based on its orbit*, how massive the satellite was. Unfortunately, they were unable to provide the U.S. president with the information he requested. However, they would have been able, instead, to tell him the mass of the Earth (which Eisenhower *wasn't* concerned about). In this project, you use a different satellite—the moon—to determine the mass of the Earth. You also explore how the scientist Cavendish performed some painstaking calculations of gravitational attraction and was able to accomplish the same thing.

What You Need

• moon

• calendar

Method

1. Determine how long it takes for the moon to circle the Earth.

2. A calendar can give a reasonable result. A more accurate value is the sidereal period, which indicates only the time it takes for the moon to circle the Earth, without consideration for how long it takes to return to a particular phase. This can be obtained from a sidereal table or by subtracting 2.2 days from the value obtained by observing the number of days from one full moon to another.

3. Calculate the velocity of the moon in its orbit based on its average radius, r, of 384,400 kilometers (3.844×10^8m). You can do this using the equation:

$$v = \frac{2\pi r}{T}$$

4. Calculate the mass of the Earth using the equation

$$M_e = \frac{v^2 r}{G}$$

where G is the constant of Universal Gravitation = 6.67×10^{-11}m^3/kgs^2

Expected Results

Using the following values:

T = 27.322 days = 2,360,621 seconds

v = $2\pi r/T$ = 1,023 meters/second

r = 3.844×10^8m

$M_e = \dfrac{v^2 r}{G}$ = 6.03×1024 kilograms

This is within 1 percent of the accepted value for the mass of the Earth of 5.97×10^{24} kilograms.

Why It Works

Newton's law of universal gravitation states there is an attractive force between any two masses in the universe. The attractive force is related to how massive the objects are and how far apart they are from each other. The gravitational force is linked to the mass and distance by a constant, "big G," called the *universal gravitation constant*. Since the gravitational force is the force that provides the centripetal force that keeps a satellite in orbit, we can solve for the mass of the Earth if we know the other variables in the equation. Similarly, knowing that the gravitational force equals the weight of an object, we can solve for the mass of the Earth.

Other Things to Try

Cavendish's famous experiment is one of our "wish-list" experiments that can be used to determine the big G and, as a result, the mass of the Earth. Gravitational force between two masses can be measured using an apparatus shown in Figures 118-1 and 118-2. The relatively small force is detected by measuring the torsion it produces in a thin filament between the masses.

The Point

The mass of a body that a satellite rotates around can be determined by the orbital period of that satellite. A key component of the force is the universal gravitational constant, G. By knowing G, it is possible to determine the mass of the Earth, using either the weight of objects on the Earth's surface or the orbital period of satellites circling around the Earth.

Figure 118-1 *Cavendish apparatus.*

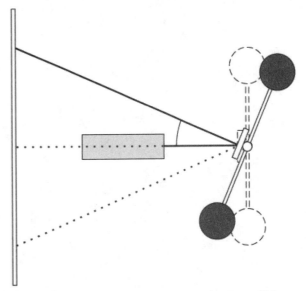

Figure 118-2 *Top view of masses in the Cavendish apparatus.*

Section 11

The Twentieth Century

Project 119
What is the size of a photon?

The Idea

One of the pivotal discoveries of the twentieth century was the recognition that light is made of *photons*, which can be thought of as minute particles of light. Light behaves in many ways like a wave, but it also behaves in many ways as if it consisted of particles. Just how big (or small) are these particles of light?

What You Need

- several LEDs (light-emitting diodes) of known wavelength
- variable power supply
- jumper wires
- voltmeter (or multimeter configured as a multimeter)
- dark room

Method

1. Attach the positive end of the voltmeter to the positive end of the power supply, and the negative end of the voltmeter to the negative end of the power supply.

2. Adjust the power supply to give a reading of zero volts.

3. Select an LED. Use jumper wires to connect the positive side of the power supply to the positive terminal of the LED. (The positive terminal of the LED is the longer one.)

4. Connect the negative side of the power supply to the LED.

5. Darken the room.

6. Slowly increase the voltage from the power supply until the LED just begins to give off light. For visible light, this will be between 1.2 volts and 2.5 volts.

7. Write down the voltage that results in light just being produced.

8. Repeat this process for all the LEDs you have.

9. Make a graph of voltage versus frequency.

The schematic for this experiment is shown in Figure 119-1.

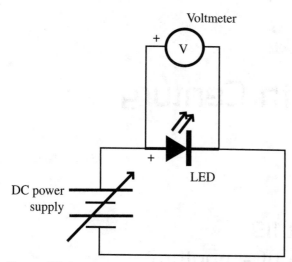

Figure 119-1 *Circuit to measure LED turn-on voltage.*

Expected Results

The higher the frequency, the higher the voltage needed to turn on the LED. The relationship is linear, as shown in Figure 119-2.

Why It Works

According to quantum theory, the energy of a photon depends on its frequency. Higher frequency light (or light closer to the blue side of the visible spectrum) has more energy. Lower frequency light (closer to the red side) has less

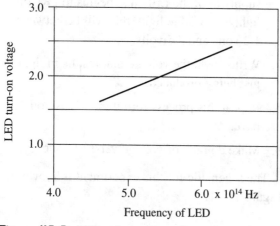

Figure 119-2 *LED voltage versus frequency.*

energy. For a photon of a given frequency, f, or color, the energy is given by hf, where h is called *Planck's constant*.

Other Things to Try

The amount of energy needed to turn on an LED is given by qV, where q is the charge of an electron $= 1.6 \times 10^{-19}$ C and V is the applied voltage. From the slope of the graph, you can estimate Planck's constant (from the slope of the previously plotted equation, $v = (h/q)f$). Planck's constant can be determined by dividing the slope of the graph by the charge of an electron. The accepted value for Planck's constant is 6.63×10^{-34} J-s. The slope of the graph in Figure 119-2 is about 7×10^{-34} J-s, which provides a reasonable order of magnitude estimate of Planck's constant.

So how small is a photon? Let's take a 60W light. This means that at about 5 percent efficiency, there are about 3 Joules of energy coming from the bulb every second. According to Planck's constant this means that every second 3 J $/6.63 \times 10^{-34}$ J-s $= 4.5 \times 10^{33}$ photons are coming from the light bulb. This incredibly large number of photons gives an idea of the extremely small size of the photon.

The Point

The turn-on voltage for an LED gives an indication of how much energy is contained in a single photon. Photons with higher frequency (shorter wavelength) have more energy than photons with lower frequency.

Project 120
How is a hydrogen atom like the New Jersey Turnpike? Seeing the energy levels of the Bohr atom.

The Idea

In this experiment, you look at the colors of the light that the atoms of a particular element give off when excited by electricity. This is the same type of data that led some of the greatest scientific minds of the twentieth century to develop the concept of the atom. The patterns of those colors give us insight into the mysteries of the structure of the atom. Like the New Jersey Turnpike, the electrons in the various energy levels of the hydrogen atom can exit only in certain specific ways.

What You Need

- diffraction grating
- tube of hydrogen
- high-voltage power supply to excite the hydrogen

Method

1. Insert the hydrogen tube into the high-voltage power supply. Make sure the good electrical contact is established between the electrodes of the hydrogen tube and the power supply.

2. CAUTION: Do *not* touch the electrical contacts of the high-voltage power supply once it is activated. *Follow all manufacturer's instructions for safe use of this equipment.*

3. Darken the room.

4. Turn on the power supply and observe a violet-blue glow in the tube.

5. Hold a diffraction grating with the scribed lines parallel to the tube in front of your eyes, as indicated in Figure 120-1.

6. Observe the image of the glowing hydrogen tube broken down by the diffraction grating. If you have a spectrometer, observe the light from the hydrogen tube and identify the positions of each of the lines you see. Look for the transmitted light to the left and right of the central image from which the glowing hydrogen tube is located. You may need to use your peripheral vision to see the entire effect.

Expected Results

The light transmitted through the diffraction grating is *not* a continuous rainbow.

The light is broken down into a few bright vertical lines.

The details of the lines you see are summarized in Table 120-1.

Why It Works

On the New Jersey Turnpike, if you get on at Exit 6 and go to Exit 7, you pay a $0.80 toll. If you go from Exit 6 to Exit 8, you pay $1.20. In a hydrogen atom, if an electron goes from the third energy level to the second energy level, only red photons (with a wavelength of 656.3 nanometers)

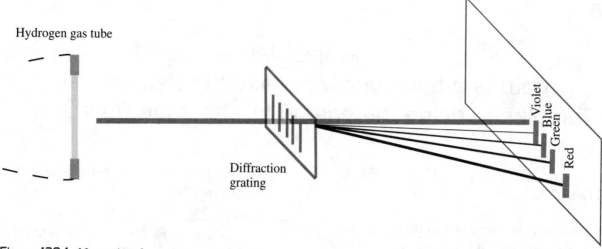

Figure 120-1 *Measuring the spectrum of the hydrogen atom.*

are released. But, if an electron goes from the fourth energy level to the second energy level, only blue-green photons (with a wavelength of 486.1 nm) are emitted.

On the New Jersey Turnpike, nothing is between Exit 6 and 7, and you never have to pay a toll between $1.20 and 0.80. The hydrogen atom does not produce a photon whose color is between red and blue-green.

Einstein's interpretation of the photoelectric effect leads us to the conclusion that photons have a certain specific energy based on their frequency. Niels Bohr developed a model of the hydrogen atom based on the idea that the electrons are found in certain specific energy levels, but not in between. A particular change in energy levels results in a photon of a particular color.

When viewed through a diffraction grating, the light from the excited hydrogen atoms does not result in a full rainbow. It, instead, produces only specific brightly colored lines, corresponding to specific wavelengths. Each wavelength is associated with a change from one energy level to another. The bigger the jump, the shorter the wavelength.

Other Things to Try

The spectral breakdown of light emitted by a hydrogen atom can also be detected using a high-sensitivity light sensor, such as PASCO part number PS-2176. The result for this is shown in Figure 120-2.

Table 120-1

Change from energy level	3	4	5	6	7	8
To energy level	2	2	2	2	2	2
Wavelength (nm)	656.3	486.1	434.1	410.2	397.0	388.9
Color	Red	Blue-green	Violet	Violet	Violet	Violet

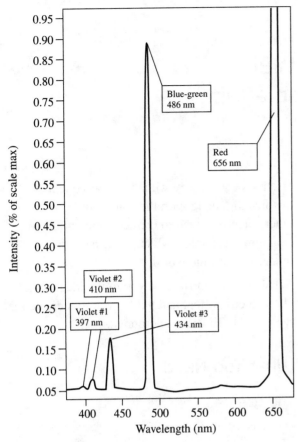

Figure 120-2 *Hydrogen emission spectrum. Courtesy PASCO.*

The Point

Observation of separate colors from glowing hydrogen gas confirms the model of the atom developed by Niels Bohr, in which electrons occupy specific energy levels. Because electrons cannot be between the established energy levels, many colors (or photon frequencies) are not produced.

Project 121
Photoelectric effect.

The Idea

In 1905, during his "miracle year," Albert Einstein published five papers. These included special relativity, which dealt with space and time, as well as general relativity, which related mass and energy through the equation $E = mc^2$. However, Einstein won his only Nobel Prize for work he did that same year on the *photoelectric effect*.

At the time, it was known that light shining on certain materials could knock out electrons to produce a current. It stood to reason that the stronger the light, the greater the current. Researchers also found that how much of a kick the electrons got (or how much kinetic energy they had) depended on the color of the light. Many scientists expected a stronger light would also release an electron with greater energy. It took Einstein's brilliance to understand why the color (or frequency) of the light played such a key role in determining how much energy the electrons came away with. The consequences of this insight, along with the contributions of many other scientists, lead to the development of quantum mechanics, which is the basis for the modern electronic world.

This project introduces you to the idea of the photoelectric effect and guides you to recreate the type of data Einstein interpreted.

What You Need

The basics

- piece of zinc metal
- sandpaper or steel wool
- short jumper wire
- source of ultraviolet light (a carbon arc lamp or possibly a strong "black light")
- source of visible light (incandescent lamp)
- plate of glass
- electroscope, either purchased or built as a project

Photoelectric effect apparatus

- photoelectric effect apparatus, such as the Daedelon EP-05 (available from www.daedelon.com)
- variable DC voltage source
- voltmeter or multimeter configured as a voltmeter
- various light sources of known frequency: this includes laser pointers of known wavelength, incandescent, carbon arc, or ultraviolet lights

Figure 121-1 *Albert Einstein explained the photoelectric by claiming that light had a particle-like nature.*

- color filters with known wavelength of transmitted light

Method

The basics

This part introduces the basic idea of the photoelectric effect and brings you to the dilemma Einstein addressed.

1. Rub the piece of zinc with a piece of sandpaper or steel wool. This removes oxides to expose the metal.

2. Discharge the electroscope by touching your finger to the electrode.

3. Using a very short jumper, attach the zinc to the electroscope.

4. Darken the room.

5. Shine the light from an ultraviolet source onto the zinc.

6. Observe the effect on the electroscope leaves.

7. Discharge the electroscope and compare the effect of the ultraviolet source and the visible source. Also compare the effect of shining the ultraviolet source through a pane of glass that transmits mostly visible range light, but hardly any ultraviolet light.

8. Charge the electroscope positively and observe the effect of shining ultraviolet light on the zinc.

9. Charge the electroscope negatively and observe the effect of shining the ultraviolet light on the zinc.

Photoelectric effect apparatus

This approach uses a metal target in a vacuum tube. Because the currents that need to be measured are so small, it is helpful to have the detector very close to the source of the current. This procedure goes through the generic steps to make this measurement with specific references to the EP-05 operation (more detailed instructions are available with that apparatus):

1. Set up the fluorescent lamp to focus on the detector (photodiode).

2. Attach a voltmeter to read the stopping voltage (stopping potential) across the photodiode. (The connections are the red and black banana jacks on the EP-05.)

3. Place the blue filter over the opening going into the photodiode. The apparatus should be set up as shown in Figure 121-2.

4. Darken the room. If necessary, construct a light shield from a cardboard box to protect the photodiode from stray light.

5. Adjust the stopping potential, so all the electrons are turned back and there is no photocurrent. (This is accomplished by turning the "voltage knob" to the full clockwise position.)

6. Now, adjust the stopping potential to its minimum value. (This can be done by turning the voltage knob as far in the counterclockwise position as possible.)

7. Adjust the radiation intensity by changing the distance between the light source and the detector to read about 10 on the intensity scale. You are now calibrated and ready to make some measurements.

8. Measure the current reading and write down the reading on the voltmeter.

9. In several steps, increase the stopping potential and record the current reading at each step. Five readings should be sufficient to define a linear relationship.

Figure 121-2 *Photoelectric effect apparatus.*

10. The data should produce a linear relationship similar to the one shown in Figure 121-2 between stopping potential and voltage.

11. The voltage required to produce zero current is a key point that determines the value of the work function for the metal (in the photodiode).

12. Repeat the previous steps using the green filter.

13. Replace the fluorescent lamp with a tungsten incandescent lamp. Install the red filter and repeat the previous steps until current versus voltage curves for the red filter.

14. A laser or LED of known wavelength can also be used as a source of illumination. A diverging lens (biconcave) may be helpful in spreading the laser beam to fill the opening area of the photo diode.

15. For each color, plot the current versus voltage and extrapolate the curve to find the threshold stopping voltage that results in zero current.

16. Plot the stopping voltage versus the frequency for each of the frequencies (colors) for which you took data.

Expected Results

Ultraviolet light shining on a piece of zinc results in a charge separation. This charge causes the leaves of a negatively charged electroscope to separate further and causes the leaves of a *positively* charged electroscope to come together. This indicates the charge is negative or, more specifically, consisting of electrons. Visible light does *not* result in this charge being developed in the zinc.

Using the photoelectric effect apparatus, we find that:

1. The greater the frequency, the greater the stopping voltage required to limit the current flow. This relationship is linear. This means the kinetic energy of the freed electrons is proportional to the frequency of the light.

2. Below a certain *threshold frequency*, no current is generated.

3. Increasing the intensity of the light increases the current (for a given stopping potential and light frequency). However, increasing the light intensity does not have any effect on the kinetic energy of the freed electrons.

4. The slope of the stopping voltage versus the frequency graph represents Planck's constant divided by the charge on one electron. The equation for this is:

$$V = \frac{hf}{q}$$

Because the wavelength of light is usually more readily available, the frequency can be determined from the equation:

frequency = speed of light / wavelength

5. From the slope of the voltage versus frequency graph, Planck's constant can be determined from the slope multiplied by the electronic charge: $q = 1.6 \times 10^{-19}$ C. A slope of $4 = 10^{-5}$ gives the expected value of Planck's constant.

Why It Works

Light contains energy based on its frequency. The frequency of visible light is lower than that of ultraviolet light and it does not have enough

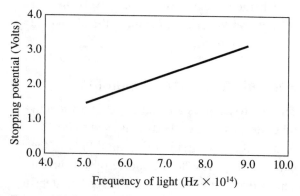

Figure 121-3 *Stopping potential for electrons exposed to various frequencies of light. The slope of this line determines Planck's constant.*

energy to free electrons from a metal, such as zinc. As the frequency of the light increases, the energy each photon carries is raised above the threshold required to free electrons from the zinc.

The *work function* of a metal is a measure of how tightly electrons are held by the atoms of the metal. If the photon energy is greater than the work function of the metal, electrons are released. If the freed electrons encounter a stopping voltage (stopping potential), the amount of extra energy above the work function can be determined.

This can be summarized by the equation:

$$KE = E_{photon} + W$$

where *KE* is the kinetic energy of the freed electron (measured by the amount of voltage required to stop the electrons).

E_{photon} is the energy carried by the photon.

W is the amount of energy just to free one electron from the metal with no extra energy to get it moving.

The energy in a photon was given by:

$$E_{photon} = hf$$

where *h* is Planck's constant and *f* is the frequency of the light.

Other Things to Try

A good software simulation of the results of this experiment can be found at http://phet-web. colorado.edu/wb-pages/simulations-base.html.

The Point

The key concept underlying this experiment is that light energy comes in specific amounts or packages called *quanta* or *photons*. These photons cannot be broken up into smaller units. The higher the frequency of the light, the greater the amount of energy contained in one photon. If a photon has enough energy to release an electron, an electric current can flow; otherwise, below that threshold, no energy will flow. The more energy the photon has, the more kinetic energy the electron processes when it is released.

Project 122
Millikan oil-drop experiment. Mystery marbles.
Understanding how the experiment worked.

The Idea

Robert Millikan devised a brilliant technique to experimentally determine the charge of the electron, which resulted in him being awarded the Nobel Prize for Physics. This project lets you replicate Millikan's famous experiment. Basically, Millikan found a way to attach electrons to small droplets of oil, and then measure their response to an electric field. Because this is a more complex experiment than most of the other experiments in this book, it may be out of reach for many readers.

For this reason another option to explore this discovery is given. One of the problems Millikan had to deal with was he never knew how many electrons were on any given drop of oil. We can re-create some of the logical steps Millikan followed using pennies to represent electrons.

What You Need

Simulation

- film canisters or plastic prescription containers with covers

- spray paint

- about 150 pennies

- digital scale or spring balance

Replicating the Millikan oil-drop experiment

- Millikan's oil-drop apparatus, as shown in Figure 122-1

Method

Setting up the simulated oil drops

1. Spray paint or otherwise obscure the outside of about 8–12 plastic containers, so you can't see inside.

2. Measure the mass of the empty containers.

3. Distribute a different (random) number of pennies in each of the containers.

4. The easy version of this includes at least one set of containers that differ by one penny (such as Container 7 with 12 pennies and Container 8 with 11 pennies).

5. A slightly more challenging version is to have no container differing by one penny, but (because of the small statistical sample) to have at least one set of samples differ by two pennies and another set by two pennies.

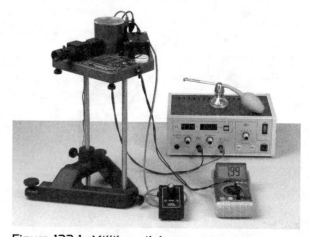

Figure 122-1 *Millikan oil drop experiment apparatus. Courtesy PASCO.*

Finding the charge of a simulated "electron" (mass of a penny)

1. Find the mass of each container with the pennies.

2. Subtract the mass of the container to obtain the mass of just the pennies in each container.

3. Arrange the mass measurements in order—smallest to largest.

4. Subtract each mass measurement from the previous measurement in the list.

5. Identify the smallest (non-zero) mass difference between any pair of containers.

6. Divide each of the mass differences by the smallest mass difference in the list.

7. If any fractional numbers are in the list, multiply all the number by a factor that leaves only integers in the list. (For instance, if one of the numbers is 1.5, multiply them all by 2.)

8. Make a graph of the mass differences on the y-axis versus the integers in Step 7.

9. Find the slope of this graph. This should give you the mass of the penny, following a similar form of logic Millikan used to measure the charge of the electron.

The actual Millikan oil-drop measurement

1. Determine the mass of the oil drop by measuring the velocity of the drop as it falls. Because air resistance affects larger drops to a greater extent, the velocity serves as a very accurate measure of the droplet mass.

2. Using X-rays or another source of ionizing radiation, create a random number of charges on the electron.

3. Determine the magnitude of the electric field that just balances the gravitational pull on that droplet. The gravitational force can be found from the mass of the droplet determined in Step 1 and the density of oil. The greater the charge, the greater the force needed to balance it.

4. At this point, we know the charge, but we don't know how many electrons are on any given droplet. This is very similar to the situation we just addressed with the pennies. Although we did not know how many pennies were in any particular container, we were able to find the mass of a single penny. Using a similar logic, Millikan was able to find the mass of an electron.

Expected Results

Following the previous simulated procedure using pennies, the slope of the line in Figure 122-2 is 2.7 grams, which is the mass of a single penny. This is a reasonable average for pennies minted before and after 1982. A more precise value can be established by sorting pennies into groups before and after 1982.

The charge of an electron determined by Millikan is -1.6×10^{-19} Coulombs.

Other Things to Try

Marbles can be used to simulate the logical process pursued by Millkan in a similar manner that was done with pennies. The marbles have a greater mass, which may make it easier to detect difference. However, finding a relationship graphically may be more difficult because of the variation in mass for a random set of marbles.

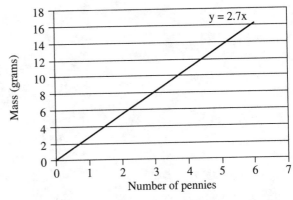

Figure 122-2 *Using the mass of a penny to simulate the photoelectric effect.*

Why It Works

The size of an oil drop is found by observing its free-fall velocity in air. The oil drop is then given a charge by exposing it to ionizing radiation. The electric field that establishes equilibrium with gravity is related to the force. Although the exact number of electrons on any give oil drop cannot be determined directly, the common multiple leads us to identify the charge of a single electron.

The Point

The Millikan oil-drop experiment determines the charge of an electron by measuring the response of an oil drop charged by electrons in an electric field.

Project 123
Ping-pong ball chain reaction.

The Idea

This is a fun and simple demonstration that will help you understand how a nuclear chain reaction occurs. It was used years ago in a Walt Disney film called *Our Friend the Atom* and appeals to all readers including the youngest and technically least sophisticated.

What You Need

- 24 spring-type mousetraps
- 49 ping-pong balls
- enclosure with at least one transparent side (a large fish tank with glass sides and a glass bottom can work well)
- optional: 2 mirrors the size of, or larger than, the side of the enclosure

Method

1. Set the traps (Figure 123-1).

2. Carefully place two ping-pong balls on each of the mousetraps (where the cheese would have gone). See Figure 123-2.

3. Lay out the mousetraps in an array that will fit into the enclosure, such as a 6 × 4 array. Obviously, you need to be extremely gentle and avoid sudden motions to prevent a premature release of the mousetrap. Any mishap will likely take other mousetraps out with it.

4. Either lower the enclosure over the mousetraps or develop a way to bring the mousetraps into the enclosure. You may need to experiment with different methods of loading the mousetraps. You may prefer to

Figure 123-1 *Each mousetrap represents a uranium atom.*

Figure 123-2 *Each ping pong ball represents a neutron.*

place the ping-pong balls after, rather than before, moving the traps. You may want to develop a wooden or foamboard template that protects the trap's trigger mechanism while you are placing the ping-pong balls or glue the traps to a board.

5. With the ping-pong ball loaded on the mousetraps in the enclosure, you are ready to initiate the chain reaction. So far, you have used 48 ping-pong balls, so one should be left. The remaining ball is the neutron that starts the chain reaction.

Expected Results

As in a nuclear-fission chain reaction, a neutron (the starter ping-pong ball) creates the first fission reaction. This event is simulated by the mousetrap releasing two additional ping-pong balls. These, in turn, potentially each release two more balls (neutrons) initiating a doubling of the available neutrons with each fission. As additional ping-pong balls are released, the rate of the reaction accelerates. This chain reaction is

Figure 123-3 *After a simulated chain reaction.*

simulated by rapidly releasing ping pong balls, which in turn releases other ping-pong balls to continue the reaction. The aftermath of this is shown in Figure 123-3.

Why It Works

Nuclear fission occurs in nature when an isotope of a nuclear material absorbs a neutron and become unstable. The nucleus splits, forming two lighter "daughter" nuclei and a spray of free neutrons that produces the cascading effect known as a *chain reaction*. There also needs to be a critical mass for this process to become self-sustaining.

Other Things to Try

It would be interesting to capture a video image of this simulated nuclear reaction and view it in slow motion.

The Point

Nuclear fission is initiated by a free neutron that causes a nucleus (such as a uranium-238 nucleus) to split and release additional neutrons. This is the basis of nuclear power, which currently provides about one-fifth of the electricity in the United States.

Project 124
The sodium doublet. Why do we think the electron has both up and down spins?

The Idea

No one has ever seen an electron spin. In fact, for that matter, no one has ever even seen an electron. Yet, we know an electron behaves as *if* it were spinning. Some of the most revealing evidence for this comes from the light that certain atoms emit when they're excited.

If some sodium chloride is exposed to a flame, the flame takes on a characteristic yellow/orange color. This is the color observed in the common flame test used in chemistry labs to identify the presence of sodium in sodium vapor street lamps. If you look at the light coming from an excited sodium atom with a spectroscope or diffraction grating, the first thing you notice is a single orange/yellow line with a wavelength between 589 and 590 nanometers.

However, on closer inspection, you notice *not one but two* orange/yellow lines. The purpose of this project is to observe these two lines, known as the *sodium doublet*, and, more importantly, to understand why they are split.

What You Need

- Bunsen burner or other flame
- concentrated sodium chloride solution
- clean nichrome wire loop (or a wooden splint)
- diffraction grating or spectroscope
- sodium vapor discharge tube with appropriate high-voltage power supply

Method

1. Use one of the previous methods to produce a light source generated by excited sodium atoms.
2. Darken the room.
3. Observe the light using a diffraction grating or a spectroscope.
4. Look carefully until you see a vertical yellow/orange line. Look closely until you notice this line is formed by two separate lines. See Figure 124-1.

Expected Results

The point of this project is to observe two separate yellow/orange lines that make up the sodium doublet.

Why It Works

When an electron goes from one energy level to a lower energy level, it gives off light. Each energy level can hold two electrons: one with spin up and the other with spin down. The electron with the spin up takes a slightly greater amount of energy to go from one energy level to another. As a result, the electrons with different spin conditions give off a slightly different color (wavelength) light.

Continuing the New Jersey Turnpike analogy (from Project 120), let's say you travel a certain distance going from Exit 7 to Exit 8. But things are slightly different if you get off at either an

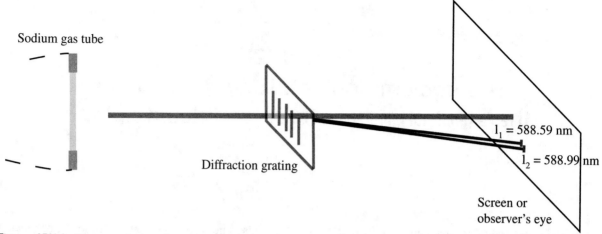

Sodium gas tube

Diffraction grating

$l_1 = 588.59$ nm

$l_2 = 588.99$ nm

Screen or observer's eye

Figure 124-1 *Electrons in a sodium atom produce a primary characteristic wavelength when electrons move from one energy level to another. A slightly different wavelength is produced depending on whether the spin of the electron is "up" or "down."*

eastbound or westbound ramp at the exit. That small difference can be thought to be something like the effect caused by electron spin.

Other Things to Try

If an excited sodium atom is exposed to a very powerful magnetic field, these spectral lines split even further. This is called the Zeeman effect, which requires magnetic fields on the order of 18 Teslas. However, because this is roughly 20 times more powerful than the very strong magnetic fields used to study nuclear magnetic resonance, we won't pursue Zeeman splitting experiments in this book.

The Point

In an atom, electrons have up or down spin. When an electron goes from one energy level to another, the energy given off by each of the two spin orientations is slightly different. Observing the split in the frequency supports the concept of electron spin.

Project 125
Building a cloud chamber. Why muons should not be here. Special relativity.

The Idea

Cosmic rays are subatomic particles that stream through the universe at high speed. As you read this, dozens of these particles are passing through your body harmlessly every second. In this project, you build a device called a cloud chamber, which will make some of these particles visible. The *cloud chamber* contains an alcohol-vapor cloud that produces vapor trails when a charged particle passes through it. This trail reveals the path or track the particles take as they pass through the vapor cloud. You begin to learn to recognize the signature of some of those particles.

Scientists believe most cosmic rays come from the sun. These cosmic rays consist of fragments of the nuclei of hydrogen or helium atoms, resulting in particles that are either singly charged (protons) or doubly charged (alpha particles). When these cosmic ray particles strike the upper atmosphere of the Earth, they collide with the air molecules present. This results in collisions that produce new particles, called *secondary cosmic rays*. These are what you will detect in your cloud chamber.

The most common result of the collision is a particle called a *muon*. The muon is a negatively charged particle, which is bigger than an electron, but smaller than a proton. The muons that are created in the upper atmosphere decay incredibly rapidly. Because of their extremely short life, they *should not be able to survive the trip through the Earth's atmosphere* to be detected on its surface. However, the most common particle

detected in the secondary cosmic ray stream is the muon. The only way to explain the abundance of muons you see in your cloud chamber is by turning to Einstein's theory of relativity, which says that time slows down for the very high-speed muons.

Also present in the shower of secondary cosmic rays hitting the Earth's surface are positrons and electrons. The *positrons* are a form of antimatter that can also be seen in your cloud chamber.

What You Need

- small 2.5 gallon fish tank
- small Styrofoam cooler
- 1 liter of *pure* isopropyl alcohol (*not 70 percent* alcohol, as is more readily available). *Pure* methyl alcohol can also be used.
- sheet of black felt large enough to line the bottom of the tank (note, the fish tank will be turned upside down, so what we refer to as the bottom of the fish tank will become the top of the cloud chamber)
- metal plate the size of the top of the fish tank—aluminum is preferable because it conducts heat better than steel
- duct tape, silicone rubber sealant, or weather stripping to make the fish tank airtight
- black (solvent resistant) paint
- about 1 pound of dry ice to encase the bottom of the fish tank
- bright light source, such as a powerful flashlight

- strong magnet

- optional: source of low-level radiation, such as a smoke detector, mantle of old Coleman lantern, or certain old ceramic objects that contain cobalt

- optional: a digital video camera

Method

Building the cloud chamber

1. Because the dry ice has a limited shelf life and, for most users, takes a special effort to obtain, a good idea is to do a dry run and assemble these parts before picking up the dry ice. Also, remember dry ice is extremely cold and should not come into contact with eyes or skin.

2. Attach the black felt to the bottom of the fish tank. Use black electrical tape or Velcro under the felt to secure it, so it remains in place when the fish tank is inverted. You may want to devise some other way to secure the felt such as small wooden supports. Keep in mind that whatever you use to support the felt must not be degraded by the solvent vapor that will be present. A solvent resistant glue such as Gorilla Glue or Gorilla tape can hold the felt without being attacked by the alcohol vapor.

3. Place the dry ice in the bottom of the cooler. If the dry ice is in large chunks, you will have to chop it. You can add a little alcohol to the mixture to form a slush to make better thermal contact with the metal plate. If you don't have a cooler the fish tank will fit in, use a box and line it with insulating material.

4. Soak the black felt with alcohol. Avoid dripping that produces puddles of alcohol on the metal plate by not using too much alcohol.

5. Cover the dry ice with the metal plate, with the side painted black facing away from the dry ice. The metal plate should be completely in contact with the dry ice. You can secure the plate to the tank first if that works better.

6. Place the tank—alcohol-soaked felt up—with the metal plate over the dry ice. It is important to establish good thermal contact between the dry ice and the metal plate. People have used solid blocks of dry ice as well as crushed ice. Good thermal contact can be achieved by creating a slurry by mixing some isopropyl alcohol in with the dry ice. If you crush the dry ice, be sure to wear protective classes to avoid extremely cold dry ice fragments from contacting your eyes. The dry ice mixed with alcohol should be around −70 degrees C or −94 degrees F.

7. Seal the chamber above the metal plate by wrapping electrical tape around the edge where the metal plate meets the fish tank. (Some people have found that pouring a small amount of isopropyl alcohol in the channel surrounding the metal plate helps form an airtight seal.) If air is drawn into the chamber, the vapor cloud may not from properly.

8. It may be necessary to keep the top of the cloud chamber (where the alcohol-saturated felt is) from getting too cold. The bottom of the chamber should be near −60 degrees C (−76 degrees F) to enable the formation of supersaturated vapor. However, the top of the chamber should be maintained close to room temperature (22 degrees C or 72 degrees F) to promote evaporation of the alcohol. To accomplish this it may be necessary to warm the top of the chamber either with your hands or with some other means to maintain the proper temperature gradient. It might be helpful to measure the temperatures of both surfaces.

9. Shine the light from the side of the tank toward the metal plate.

10. The stack should look like this, from top to bottom:

    - Bottom of the fish tank

    - Black felt soaked with isopropyl alcohol

    - Fish tank (metal plate covering the top of the tank)

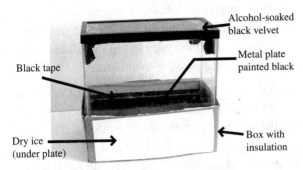

Black tape

Alcohol-soaked
black velvet

Metal plate
painted black

Dry ice
(under plate)

Box with
insulation

Figure 125-1 *Cloud chamber assembled.*

- Metal plate
- Sheets of dry ice
- Bottom of the Styrofoam cooler

Observing tracks

1. At first, you will notice a mist of alcohol droplets forming in the tank.

2. After about 15 minutes, you should start to see the tracks of particles passing through the vapor a few centimeters above the base plate.

3. It may be helpful to view the tracks by looking *toward* the light source at an angle so that the vapor trails are illuminated from *behind*.

4. If you have a low-level radiation source (such as one of the everyday objects mentioned in the parts list), place it near the edge of the cloud chamber and compare its effects to cosmic rays. (Smoke detectors have low-level radioactive materials, such as americium, that are packaged safety for its intended use. Do not attempt to dismantle a smoke detector to get at the radioactive isotope. The mantle for at least some Coleman gas-camper lanterns contains traces of radioactive Thorium, which can also be a safe low-level source of charged particles to view.)

5. Observe what a magnet does to the tracks. Note in particular how the particle is diverted in relation to the particle's original velocity and the north-to-south direction of the magnetic field.

6. Once you get this going, a video camera can be very helpful in recording the tracks and providing an opportunity to analyze the tracks in detail. Still photography is very difficult because of the randomness of the way the tracks are created and the rapidity with which they fade. Extracting still images from a video recording is more likely to produce clear images of tracks.

Expected Results

After the mist forms into a supersaturated alcohol vapor, you may start to notice tracks that look like spider webs along the chamber bottom. These are cosmic rays and should be noticeable roughly several times each minute.

Alpha particles, which are two protons and two neutrons bonded together, form sharp, well-defined tracks about 1 centimeter long.

Beta particles, which consist of electrons, have thinner and longer tracks, roughly 3 to 10 centimeters in length.

Some of the tracks may come in straight and then sharply break in a different direction. An example of this is shown in Figure 125-2, which shows a muon being deflected as it dislodges an electron from an air molecule.

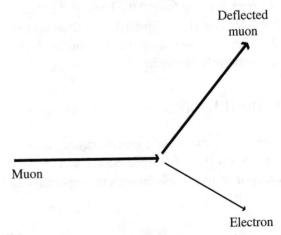

Deflected
muon

Muon

Electron

Figure 125-2 *Collision between a muon and an atom. The muon is deflected and the electron is knocked out of the atom, leaving a second, fainter track.*

A track that starts in a straight line, but then breaks off at a sharp angle, such as shown in Figure 125-3, most likely is muon decay during which a muon spontaneously decays to form an electron. The electron is visible as a thinner track. The two neutrinos do not form vapor trails and are not visible because they are not charged.

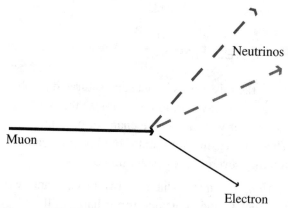

Figure 125-3 *Muon decay. A muon spontaneously decays into an electron and two neutrinos. The neutrinos do not leave tracks in the cloud chamber.*

You may see a very jagged, erratic path representing a low-energy particle being scattered multiple times. This is pictured in Figure 125-4.

If you view the chamber from the front, with the particles coming from the left, and the magnet's north pole at the top of the chamber, particles bending toward the back of the chamber are positively charged particles (such as protons). Particles bending toward the front of the chamber are negatively charged.

Why It Works

Some of the radiation in cosmic rays or isotope sources consists of charged particles. As these charged particles pass through the supersaturated alcohol vapor in the chamber, the particle ionizes the molecules of the vapor. Droplets of the vapor then condense on the path left in the wake of the particle's path, leaving a visible trail.

Collisions occur that either change the motion of the particle or result in a subatomic event, which results in a whole new mix of particles. The laws of conservation of momentum and mass must be followed, which helps to identify the particles, including those that are not visible. Nonionized particles, such as neutrons, will not leave a visible track.

Figure 125-5 shows a sampling of particle collisions in a more elaborate detection system, called a *bubble chamber*.

Other Things to Try

A muon is one of many types of subatomic particles that may be detected by your cloud chamber. A muon has the same charge as an electron, but it is 207 times more massive. Muons are created when cosmic rays strike the upper layers of the Earth's atmosphere. Muons are unstable and disintegrate into other particles almost immediately.

After being created at the top of the atmosphere, muons decay within 2.2 microseconds or 0.0000022 second. A *microsecond* is one-millionth of a second. In this time, the muon would travel only 659 meters (0.659 km). Because they are

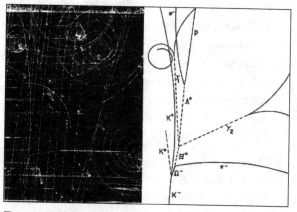

Figure 125-5 *Subatomic particle tracks. Courtesy Brookhaven National Labs.*

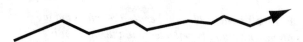

Figure 125-4 *Zigzag path of slow moving low-energy charged particles.*

created between 10 and 15 km above the surface of the Earth, most muons would decay before reaching the surface of the Earth. The trouble is how is it even possible that muons are able to travel the distance from where they are created at the top of the atmosphere to the ground before decaying? Even traveling at over 99 percent of the speed of light, the muons do not have enough time to make it to the ground. We should not see any muons at all.

However, because the muons are traveling so fast—according to Einstein's theory of relativity—time slows down for them. The fact that we can observe muons at the surface of the Earth serves as proof of Einstein's theory, which states that the time for the moving particle is

$$t = \frac{t_o}{\sqrt{1 - \frac{v^2}{c^2}}}$$

where t is the time the muon takes as observed from the Earth, t_o is the time it would take for a stationary muon, v is the speed of the particle which, in this case, is 0.99c, and c is the speed of light. When viewed through the perspective of Einstein's theory, the muon's lifetime becomes larger (35 microseconds), which gives it enough time to make it through the Earth's atmosphere before decaying.

Another track you may see is that of a positron, which is more difficult to distinguish from other positively charged particles. However, just knowing that many of the cloud chamber events are positrons is significant in itself. A positron is the antimatter version of an electron. Now, before you dismiss this as a far-fetched contribution from science fiction, the positron is very commonly found in collision fragments from cosmic rays. (By the way, if you are a science-fiction fan, antimatter plays a prominent role in Dan Brown's *Angels and Demons*.) When a positron collides with an electron, the two annihilate each other and release energy. Although it may be difficult to identify this event, your cloud chamber positron-electron annihilations are common. This subatomic particle process has actually been developed into a useful application in the form of the PET scanners found in many medical imaging labs. The P in PET stands for "positron." Today, the PET scans enable medical researchers and diagnosticians to image functions, such as the metabolism of malignant tumors and the early diagnosis of Alzheimer's disease, and to identify risk factors for heart conditions.

Although the technological applications of positron annihilation are complex, they are the same events you can observe taking place in your cloud chamber.

(Background on subatomic particles and their detection was derived from the following source, which is recommended for further information on this topic: "Cloud Chambers and Cosmic Rays, A Lesson Plan and Laboratory Activity for the High School Science Classroom," Cornell University, Laboratory for Elementary Particle Physics, 2006, available from http://www.lepp.cornell.edu/ Education/rsrc/LEPP/Education/Teacher Resources/cloudchamber.pdf.)

The Point

Cosmic rays, consisting of charged subatomic particles, are continuously striking the Earth's surface. These particles can be detected by observing the tracks they leave in a supersaturated vapor in a device called a cloud chamber.

Appendix A

Where to Get Stuff

PASCO Scientific
10101 Foothills Blvd.
Roseville, CA 95747-7100
1-800-772-8700
www.pasco.com

Sargent-Welch
P.O. Box 4130
Buffalo, NY 14217
1-800-727-4368
www.sergentwelch.com

Flinn Scientific, Inc.
P.O. Box 219
Batavia, IL 60519
1-800-452-1261
www.flinnsci.com

Frey Scientific
c/o School Specialty Science
80 Northwest Blvd.
Nashua, NH 03063
1-800-225-3739
www.freyscientific.com

Edmund Scientific
60 Pearce Ave.
Tonawanda, NY 14150
1-800-728-6999
www.scientificsonline.com

Daedalon Corporation
P.O. Box 727
Waldoboro, Maine 04572
1-800-299-5469
www.daedalon.com

RadioShack
www.radioshack.com

Appendix B

(More Than) Enough Physics to Get By. (Highly Optional)

Equations

Project 1 $v = \Delta d / \Delta t$

Project 2 $d = d_o + vt$

Project 3 $d = d_o + v(t - t_o)$

Project 5 $F = ma$

Project 8 $v_f = \dfrac{2d}{t}$

                 $t = \sqrt{\dfrac{2h}{g}}$

Project 9 $v = R / t$

                 $R = (v^2/g)\sin 2\theta$

                 $h = (v \sin \theta)^2/2g$

Project 10 $v_x = R/t$

                 $v_y = \dfrac{gt}{2}$

                 $v = \sqrt{v_x^2 + v_y^2}$

Project 13 $F_c = \dfrac{mv^2}{r}$

Project 17 $v = \sqrt{rg}$

Project 19 $g = 2d/t^2$

                 $a = 2d/t^2$

Project 22 $T = 2\pi \sqrt{\dfrac{L}{g}}$

Project 31 $a = \dfrac{(m1 - m2)g}{(m1 + m2)}$

Project 51 $\tfrac{1}{2} mv_a^2 = \tfrac{1}{2} mv_b^2 + \tfrac{1}{2} mv_c^2$

Project 56 $v = \left(\dfrac{m + M}{m}\right) \sqrt{2gh}$

Project 67 $T = 2\pi \sqrt{\dfrac{m}{k}}$

Project 68 $g = \dfrac{4\pi^2 L}{T^2}$

Project 70 $f = v/2(L + 0.8d)$

Project 77 $v = \lambda f$

                 $v = 331 + 0.6T$

Project 81 $I = I_o/r^2$

Project 83 $\lambda = \dfrac{d \sin\theta}{n}$

Project 85 $n_i \sin (\theta_i) = n_r \sin (\theta_r)$

Project 86 $I/I_o = \cos^2\theta$

Project 96 $F = \dfrac{kq_1 q_2}{r^2}$

                 where k is the Coulomb constant $=$ 9.0×10^9 m²/C²

Project 100 Ohm's law
                 $V = RI$
                 $R = VI$, and
                 $I = V/R$

                 $R_{series} = R_1 + R_2$

                 $1/R_{parallel} = 1/R_1 + 1/R_2 + \ldots$

Project 103 The current for a charging capacitor is given by $I = I_o e^{-t/RC}$

                 The voltage for a charging capacitor is given by $V = V_o(1 - e^{-t/RC})$

The current for a discharging capacitor is given by $I = I_0 e^{-t/RC}$

The voltage for a discharging capacitor is given by $V = V_0 e^{-t/RC}$

Project 116 $I = I_0 e^{qV/kT}$

Project 121 $E_{photon} = hf$

Project 125 $t = \dfrac{t_0}{\sqrt{1 - \dfrac{v^2}{c^2}}}$

Units used in this book

Length

1 m = 3.28 ft

1 mile = 1.61 km

1 inch = 2.54 cm

Time

1 day = 86,400 s

1 year (365¼ days) = 3.16×10^7 s

Energy and Power

1 J = 0.738 ft-lb = 0.239 cal

1 kW-hr = 3.6×10^6 J

1 W = 1 J/s

1 hp = 746 J

Force

1 N = 0.225 lb

1 lb = 4.45 N

Weight and mass

A 1 kg mass has a weight of 9.8 N (on Earth)

A 1 kg mass has a weight of 2.2 lbs (on Earth)

Speed/velocity

1 m/s = 2.24 mi/hr

1 km/hr = 0.621 mi/hr

1 km/hr = 0.278 m/s = 0.91 ft/s

Volume

1 cc = 1 cm^3 = 1 mL

1 liter (l) = 1000 cm^3

1 gal = 3.79 liters

Pressure

1 Pa = 1 N/m^2 = 1.45×10^{-4} lb/in^2

1 atm = 1.01×10^5 Pa

   = 14.7 lb / in^2 (psi)

   = 760 mm-Hg

   = 760 torr

Examples of prefixes

1 km = 1000 m

1 kg = 1000 g

1 m = 100 cm = 1000 mm

1 liter = 1000 mL

1 μm = 1×10^{-6} meter

1 nm = 1×10^{-9} meter

1 megohm = 1×10^6 ohm

Index

Index